The Complete

VLF-TR
METAL DETECTOR HANDBOOK

Ram Publications

Complete Book of Competition Treasure Hunting (The)
Learn the INSIDE FACTS and you, too, can become a WINNER.

Complete VLF-TR Metal Detector Handbook (The)
THE OPERATIONAL/TECHNICAL MANUAL ... thoroughly explains VLF/TR metal/mineral detectors and HOW TO USE them. Compares VLF/TR's with all other types.

Detector Owner's Field Manual
The world's most complete field guide. Explains the total capabilities and HOW TO USE procedures of all types of metal detectors.

Electronic Prospecting
Learn how to find gold and silver veins, pockets, and nuggets using easy electronic metal detector methods.

Gold Panning Is Easy
This excellent field guide shows you how to FIND and PAN gold as quickly and easily as a professional.

"How to Test" Detector Field Guide
Learn how to find QUALITY before you buy ... BFO, TR, VLF/TR, and discriminators.

Journals of El Dorado (The)
The most misunderstood treasure hunting book. An invaluable research tool. Study this book and FIND TREASURE.

Professional Treasure Hunter
Discover how to succeed with PROFESSIONAL METHODS, PERSISTENCE, and HARD WORK.

Successful Coin Hunting.
The world's most authoritative guide to FINDING VALUABLE COINS with all types of metal detectors. The name speaks for itself!

Treasure Hunter's Manual #6
Quickly guides the inexperienced beginner through the mysteries of FULL TIME TREASURE HUNTING.

Treasure Hunter's Manual #7
The classic! THE book on professional methods of RESEARCH, RECOVERY, and DISPOSITION of treasures found.

Treasure Hunting Pays Off!
An excellent introduction to all facets of treasure hunting.

ON THE FRONT COVER: The instruments displayed represent a few of the many VLF-TR discriminator models that are produced by the majority of the manufacturers of VLF-TR discriminating metal detectors.

The Complete

METAL DETECTOR HANDBOOK

By
ROY LAGAL and
CHARLES GARRETT

ISBN 0-915920-32-8
Library of Congress Catalog Card No. 78-60309
The Complete VLF-TR Metal Detector Handbook
©Copyright 1979. © Copyright 1980. Roy Lagal and Charles Garrett.
First printing June 1979. Third printing May 1980.

PORTIONS OF THIS BOOK MAY BE REPRODUCED. All rights reserved, except permission is granted to reproduce printed text and/or illustrations, not to exceed the equivalent of one chapter, provided the book title is included, full credit is given to the author, and the name and address of the publisher are given. For use of matter to exceed that amount, or to reproduce the book in series form, it will be necessary to secure specific written permission from the publisher. Address all inquiries to Ram Publishing Company.

Printed in U.S.A. by
Yaquinto Printing Co., Inc. • 4809 S. Westmoreland • Dallas, Texas 75237

For FREE listing of related treasure hunting books write
Ram Publishing Company • P. O. Drawer 38649 • Dallas, Texas 75238

We dedicate this book to our wives, Gerri and Eleanor, who have been patient, understanding, and helpful over the years while we spent countless hours in the field, testing instruments and gathering data so that this book could be written. Behind many successful endeavors and works of note there is often a woman who does not receive just credit for her valiant support.

Contents

From the Authors *page xi*

Part 1 Introducing the VLF

1. History 3
2. Development 9
3. Theory 21

Part 2 Understanding the Operation and Capabilities of the VLF and Other Detector Types

4. Operating Characteristics of the VLF Metal/Mineral Detector 39
5. Comparing the VLF with the Standard TR 46
6. Comparing the VLF with the PI 49
7. Comparing the VLF with the RF, Two-Box 52
8. Comparing the VLF with the PRG 55
9. Comparing the VLF with the BFO 56
10. Comparing the VLF with Discriminators 59

Part 3 Using the VLF and Other Detector Types in All Phases of Treasure Hunting

11. Searching for Money Caches 67
12. Searching for Coins 88
13. Searching the Ghost Towns 109
14. Prospecting 123
15. Rocks, Gems, and Minerals 150
16. Gold Panning and Metal Detectors 154
17. Relic Hunting 159
18. More Detector Operating Tips 165
    Appendix A: Detector Manufacturers 177
    Appendix B: Dredging and Prospecting Equipment 178
    Appendix C: Magazines and Periodicals 179
    Appendix D: Treasure Hunting and Prospecting Schools 179
    Appendix E: Metal Detector Dealers 180
    Recommended Supplementary Books 184

Author ROY LAGAL knows more about prospecting with the aid of electronic metal/mineral detectors than any other man. His knowledge and expertise are based upon a lifetime of treasure hunting and prospecting with metal detectors. Roy and his wife traveled for more than twenty years in the gold fields, including Canada and Mexico. During the past ten years he has worked extensively with Charles Garrett and others to develop the field of electronic prospecting to a fine art. Roy, author of four books, including *Detector Owner's Field Manual* and co-author of two others, is the designer of the "Gravity Trap" gold pan. Roy gives unselfishly of his time and knowledge about prospecting and treasure hunting to all who ask.

Author CHARLES GARRETT, President of Garrett Electronics, Inc., has been treasure hunting and prospecting since his teenage years. He has traveled in Australia, Canada, Europe, Mexico, and the United States in his search for treasure and precise personal knowledge of detector applications in order to perfect not only his treasure hunting techniques but the design of the metal detectors that he manufactures. Charles has been writing for the treasure hunting and prospecting field since the late 1960s. In addition to numerous articles, he has written several books, including the popular *Successful Coin Hunting* and *Electronic Prospecting,* the latter co-authored with Roy Lagal.

From the Authors

Considering the strong hold the new VLF/TR discriminating metal/mineral detectors have gained on the metal detecting market, we realized that an applications/technical handbook, one that told the VLF story from start to present, was needed. Many erroneous beliefs regarding the history and development of the VLF's need to be corrected. Many VLF operational characteristics and capabilities had to be explained. Further, we knew that the VLF user of today and tomorrow needed all the applications information he or she could get in order to utilize the new VLF's to their fullest.

We had many lengthy discussions as to the scope and nature of a book. How detailed should it be? For instance, how thoroughly should VLF theory be presented? Should the book describe VLF operation only or should the field applications section also include all the popular detector types, such as the TR and BFO? As you study through this book, you will see the conclusions we reached. You will see that we decided to include the TR, the BFO, and other types in the various chapters. We believed that since you probably are already familiar with the other types of detectors, it would be helpful to your understanding of the VLF's if we compared them with other types when possible. We thought, too, since many of you own one or more of the other types, that this review information would strengthen your detector knowledge and perhaps make your future searches much more fruitful, regardless of the type of detector used.

In certain areas we extracted information from Roy's *Detector Owner's Field Manual* and from Charles' *Successful Coin Hunting* and other books and publications. Field information about coin and cache hunting, prospecting, etc., is as true today as it was yesterday. To it we added the "how to" VLF information we have gained from field trips all over the world.

We make no mention of brand names or manufacturers' particular products as this is an applications and technical manual explaining the history, theory, development, capability, applications, and results obtainable from the very low frequency (VLF) principle of detector operation.

The authors are greatly indebted to Editor Bettye Nelson, Ram Publishing Company. This gracious and patient lady spent many often frustrating months editing our book. She removed countless mistakes from our hand-written field notes and turned *The Complete VLF/TR Metal Detector Handbook* into a publication that will, we believe, become the standard reference on metal/mineral detectors and their applications.

Our thanks, also, to Robert Podhrasky, Chief Engineer, Garrett Electronics, for preparing the detailed VLF electronic analysis that you will find herein.

Preparing this book for you was a joy and a rewarding experience. We hope you will make every effort to apply the principles set forth in these pages. We believe that if you do, your rewards will be great. Happy Hunting ... and God bless you all!

Part I

Introducing the VLF

CHAPTER 1

History

A VLF ground canceling/discriminating metal/mineral detector is an electronic device that operates in the very low radio frequency spectrum. Its circuitry "cancels" or inhibits the effects of the earth's iron minerals upon the device while not inhibiting the detection of metal. The detection of metals is the primary design purpose of the device; the cancellation of the earth's iron minerals is secondary. These devices or instruments are known by various name descriptions: VLF — Very Low Frequency; VLF/TR — Composite Circuit VLF and TR Discrimination; GEB — Ground Exclusion Balance; MF — Mineral Free; GCD — Ground Canceling Detector; Magnum; and others.

Since man turned on the first metal detector, certain minerals present in the earth's outer surface have plagued metal detector operation. Not all of the earth's surface minerals are a bother, only two: iron and salt. Salt presents no problem except when it is wet, as found on the beaches, for example. It then becomes electrically conductive and appears to the metal detector just as if it were conductive metal. The majority of today's metal detectors, including the VLF, cannot distinguish between wetted earth salt and metal, nor can these metal detectors cancel or nullify the troublesome salt.

On the other hand, however, for the most part the effect of iron minerals can be canceled by a detector without the detector's losing the capability to detect metal. This electronic fact has saved the lives of countless people. Millions of metallic mines have been buried at shallow depths on thousands of battlefields. Since World War II, metal detectors have been employed to find those mines. Mine detectors that have the capability of canceling iron ground minerals do a more efficient job than those which cannot cancel iron ground minerals because the earth's iron minerals have a masking effect upon the response of non-ground canceling detectors, presenting problems that reduce the ability of the operator to find the mines.

3

The VLF type ground canceling metal detector is not a new idea. In fact, a ground canceling detector was developed by Hazeltine Service Corporation under a National Defense Research Committee contract in 1942. The system worked quite well and several models were built and tested.[1]

Since then, a number of "mine detectors" have appeared on the market by way of the Army surplus stores and from at least one manufacturer. Similar metal detection devices were also offered by other manufacturers. One example is a 1960s model described in *Electronic Metal Detector Handbook* by E.S. LeGaye.[2] Another is a detector which Plessey Company, Ltd. in England offered for sale in the early 1970s. They called it a "Continuous Wave Eddy Current Detector." It employed phase sensitive detectors to provide information about the type of material detected as well as to eliminate unwanted ground iron mineral responses.

Another application of this type detector was used by the Outokumpu Oy company of Finland. Their system was used in processing plants to detect conductive objects such as drill bits, bolts, and dipper teeth in the iron ore being processed. This system passed the ore or earth material over the searchcoils. Today, we pass our hobby type searchcoils over the earth. In the U.S. this same English and Finnish technology was later developed into walk-through detectors employed for airport security purposes.

Non-destructive testing of metals is another, related field where these phase sensitive systems are applied. The problem is to measure resistive and reactive properties of material to evaluate composition and to detect flaws in given materials. Hugo L. Libby and R. C. Hentschel are among the people who have been granted patents for such equipment.

So it was that the first ground canceling metal detectors were Army mine detectors. When mine detector circuits are

---

1. Office of Scientific Research and Development, *Detection of Land Mines and Sound Ranging* (Washington, D.C., 1946), pp. 13-16.

2. E.S. LeGaye, *Electronic Metal Detector Handbook* (Houston: Western Heritage Press, 1969), p. 95.

analyzed, it is learned that the VLF ground canceling metal detectors used by today's treasure hunters utilize the same natural laws and the same electronic theory. After all, the electrical properties of earth minerals and metals do not change with age; they are the same today as they have been during the countless eons of past earth history. The basic electronic circuit functions used in today's treasure hunting instruments are the same as in the Army's electronic circuits; that is, they cancel or reduce the earth's iron mineral electronic signal component while letting metallic electronic signal components "come on through"!

From the time Hazeltine demonstrated that ground canceling was possible (see Footnote 1), the Armed Services continued investigation and improvement of ground canceling circuits. Following the development of transistors and solid state multi-circuit technology in the 1950s, great strides were made in this area of electronics. Since the early 1960s, a great deal of research has been done and articles have appeared describing various techniques for achieving ground cancellation and identification of metals. Many of these methods and circuitry ideas have been incorporated into ground canceling and discriminating detectors manufactured for use by treasure hunters, coin hunters, prospectors, relic hunters, rock hounds, archaeologists, police, surveyors, plumbers, utility and electrical companies, as well as by many industries. Actually, the VLF ground canceling detectors are transmitter-receiver (TR) detectors that operate at much lower frequencies than is usual for TR types. It is for this reason that VLF detectors are often referred to as TR type detectors.

It is unfortunate that Earl Farris of the I.G.W.T. (In God We Trust) metal detector company of New Mexico did not live to complete the development of his metal detectors. If he had, there is a good possibility that treasure hunters would have had lightweight, simplified, commercial ground canceling detectors at their disposal six to eight years earlier than they did. He had described and developed detectors that canceled iron earth minerals. There are rumors that one or more persons, working the late 'sixties, continued Earl's work and/or combined it with Army-type circuits and built ground canceling instruments which they used and sold on a very small scale. Since the early

Charles Garrett holds two large nuggets which were found with VLF/TR metal detectors in the Western Australian outback gold fields. Millions of dollars in gold are being found throughout the world by persons who are using modern detection equipment. The nugget on the reader's left is a seven-pound gold nugget in a quartz matrix. The gold is dispersed throughout the entire nugget and the gold content is estimated to be at least three-fourths of the weight of the nugget. The nugget on the right weighs twenty-two ounces. The total combined value placed on the two nuggets is $100,000.

to mid-1970s, several American and English companies began to produce their versions of VLF ground canceling detectors.

The future of VLF/TR detectors looks bright, for it appears they will continue to capture a large share of the market for some time. Certainly coin and treasure hunters are finding countless older and more rare coins, as well as deeper money caches, than was ever possible with most other detector types. Nugget hunting and prospecting with certain of the specially designed VLF types can be many times more highly successful pursuits than is possible with the BFO units which in the past were the prospector's mainstay.

Combination circuit (VLF and TR) detectors are the most versatile of the ground canceling instruments available today. Almost flawless rejection performance (discrimination) can be achieved with little or no loss in depth. Even mineralized rocks and pebbles which have plagued VLF operation can be rejected

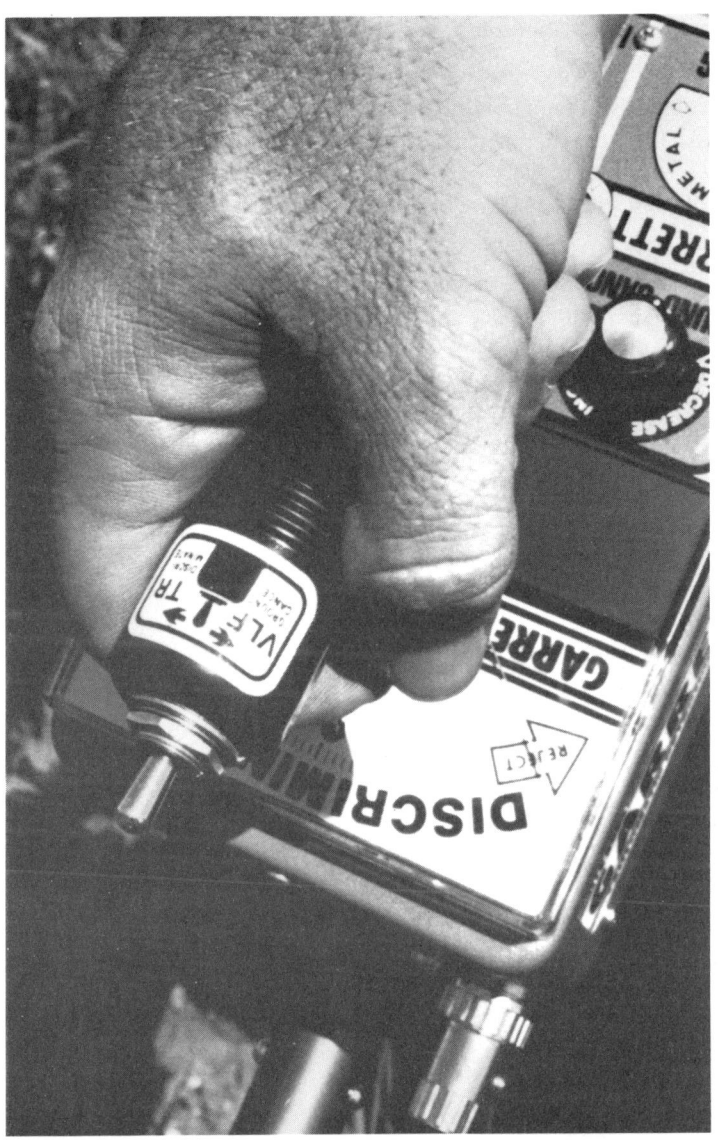

One noteworthy mode selection/tuning advancement is this master control switch. The center-position toggle switch permits the operator to change the detector's mode and accomplish retuning instantly, all at the same time. By pressing the switch to the left and holding for approximately one second, the detector electronically shifts into the VLF ground canceling mode and retunes itself. By pressing the switch momentarily to the right, the detector instantly and electronically shifts into the TR discriminating mode and the circuit is automatically retuned. This master control switch operates extremely fast and accurately.

with some brands of these combination circuit instruments.

A person who has not tried one of the new VLF/TR's is certainly in for a surprise! Though it might be difficult at first to switch from the tried and proved BFO, the veritable workhorse of the industry, or to retire that trusty TR which has filled pocket after pocket with coins... the VLF detectors are a step into the future, a step for the better.

One thing to consider is that the good hunting places where the BFO and TR are most effective are becoming much more difficult to find. Ground canceling detectors change all that. Hunting sites which have been worked and re-worked with the old type detectors still contain deep treasures which can be found by those who use the new VLF superdepth instruments.

CHAPTER 2

Development

Many major improvements have been made since the time when the first treasure hunting, ground canceling detectors were built and sold on a major scale. The first instruments utilized an operating frequency of approximately 1 kHz (kiloHertz or cycles per second) which was very close to the Army mine detector frequency. Since then, it has been found that higher frequencies produce excellent results, not only in depth capabilities but also in operation in the TR discrimination mode over highly iron mineralized ground. Most ground canceling detectors operate in the frequency range of 5 kHz to 20 kHz, as we will describe in the next paragraph.

All electrical or electronic equipment operates at signal frequencies from zero frequency (direct current — DC) up to many, many millions of Hertz (cycles) per second. The entire range of electrical frequencies is classified into many frequency bands or "groups" of frequencies. For instance, we hear frequency sounds in the 20 Hertz to 20,000 Hertz band. Within this audio band there are three smaller band classifications. The ELF (Extremely Low Frequency) band includes all frequencies from 30 Hertz to 300 Hertz. The VF band (Voice Frequency which does not include the very low and very high music frequencies) includes all frequencies from 300 Hertz to 3,000 Hertz. The VLF (Very Low Frequency) band includes all frequencies from 3,000 Hertz to 30,000 Hertz.

Since the majority of manufacturers today have designed ground canceling detectors to operate at a frequency somewhere between 5,000 Hertz and 20,000 Hertz, which obviously places them within the VLF frequency band, they were named VLF type instruments.

It was soon learned that it was important to build the VLF detectors with a wide tuning range. The iron mineral composition of the soil (matrix) which is found throughout the world varies greatly. In the eastern states iron minerals are generally

Roy Lagal scans a mine tunnel with a VLF/TR type detector. The mine had been a silver producer and contained conductive ore. While only a few specimens were discovered in the network of tunnels, they were of an extremely high grade and well worth the effort. There are countless thousands of such abandoned mines throughout the world and the modern electronic prospector can reap a gold and silver harvest from them. Only a small expense is involved and very little hard labor ... just patient searching with a metal detector.

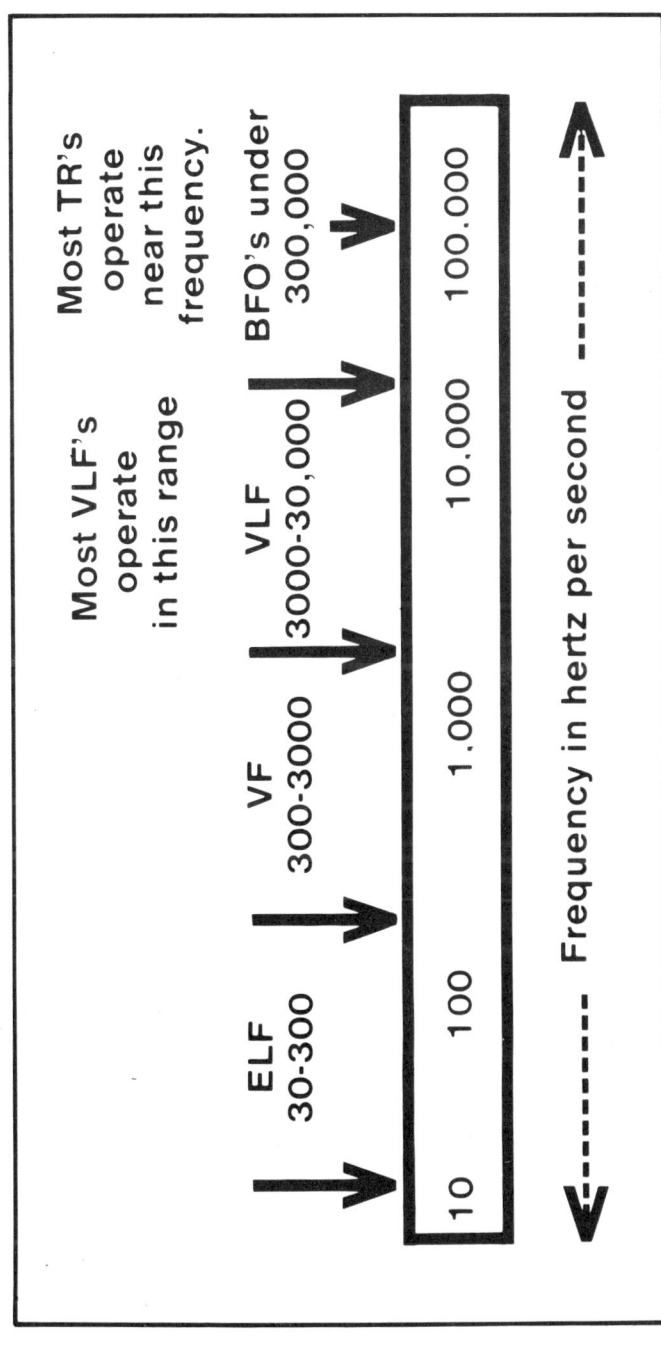

Fig. 1. Frequency chart showing the portion of the spectrum in which metal detectors operate.

It takes a great deal of time and effort to produce sleek, well designed, well balanced instruments with the control knobs engineered and located in such a way as to produce what is called "human engineering." Controls must be placed where they can be best utilized by the operator. The main controls that are subject to being bumped by brush are protected by features like handles or are placed in positions where they will receive the most protection. Our hats are off to all manufacturers who can continue to meet today's consumer demands by producing well designed, rugged instrumentation.

less troublesome than in a few of the western states, such as Washington, Idaho, California, and Colorado. Some of the toughest iron mineral concentrations in the world are found in Idaho, Washington, and Australia. All manufacturers should test their instruments in certain areas of the world where even the best of VLF detectors become somewhat difficult to operate and sensitivity begins to suffer. If their instruments operate satisfactorily in those areas, they can be assured that there will be few, if any, areas of the world where their instruments will not work.

The first VLF instruments were manufactured with one mode only... ground canceling operation. It was quickly discovered that a second (and most important) mode of operation, TR discrimination, could be incorporated. With a switch, the opera-

Charles Garrett searches in Washington state for rich gold stringers that might have been overlooked by the original miners. The area is rich in crystalline gold pockets. One small specimen can be worth several thousand dollars. Charles and Roy display several crystalline specimens in their collection at gold shows. Crystalline gold is one of the most beautiful types of gold on earth, as well as being one of the most rare.

tor could then select *either* the VLF ground canceling mode or the TR discrimination mode. The very low frequency of operation was not particularly suitable, however, especially in the 1 kHz range. It became obvious that higher frequencies would have to be used. It was found that frequencies in the 5 kHz range produce super VLF depth so necessary for the professional treasure hunter and frequencies in the 15 kHz range produce excellent depth *and* ease of operations in the coin hunting TR discrimination mode.

Thus, basically, two classes of VLF's emerged. Consider the quality professional instruments that exist on today's market. There are the deep seeking types that operate near 5 kHz. These detectors excel in depth capability while operating in the VLF ground canceling mode, whether or not the ground is heavily mineralized. Large single coins and Civil War Minie balls can be detected to depths of more than eighteen inches. Quart fruit jars filled with coins can be detected to depths of three and four feet. Operation in the TR discriminate mode over highly mineralized ground, however, presents problems. For this reason, the most popular coin hunting instruments are those that operate near the 15 kHz frequency. TR discrimination mode operation at this frequency, even in highly mineralized ground, is very good and good targets, such as coins, rings, nuggets, brass buckles and buttons, and bullets, can be quite readily distinguished from junk targets, such as bottlecaps, small iron pieces, pulltabs, etc.

In areas that contain little or no mineralization, any of the VLF's operate perfectly in the TR discrimination mode and the lower frequency models will produce coins and similar items at extreme depths.

There is, as yet, no detector built that discriminates perfectly while operating in the VLF ground canceling mode. There are several manufacturers producing VLF discriminators that require the detector to be operated in some unusual manner. Attempting to discriminate in the VLF ground canceling mode may cause certain undesirable results. At certain depths good targets may be ignored and junk targets accepted. Rejection of bottlecaps and other undesirable targets may be impossible.

Charles Garrett points to a rich stringer of high grade silver ore. This vein gave a definite signal (positive) of a conductive target, but some veins contain iron mineral and consequently (depending upon amounts of silver and iron) readings indicative of a conductive target may or may not be produced by the detector. It pays to investigate all detector responses thoroughly, especially when you are in an area where high grade native ore is known to occur frequently.

High weeds are no problem for correctly designed, electrostatically shielded searchcoils. Even in areas where rocks and heavy growth prevent searchcoils from being scanned close to the ground, today's VLF/TR detectors produce the added depth necessary to allow operators to scan with their searchcoils several inches or even a foot or more off the ground and still achieve very respectable depth detection.

The VLF searchcoils available are about as many and varied as the number of manufacturers who produce them. The most popular searchcoils are of the co-planar type. They range in size from about six inches up to about sixteen inches. The most popular sizes are approximately seven to eight inches in diameter. This size range produces excellent small object detection (nuggets, coins, rings, etc.) and will also detect quite deeply for larger objects (money jars, and so on). The authors' research proved that the seven to eight inch diameter searchcoils produce the best all around results in coin and small object hunting. A much wider scan area can be covered in a single sweep, however, with larger searchcoils and sensitivity on small nuggets is not necessarily lost. Seven to eight inch diameter searchcoils will detect coins and nuggets to greater depths than will smaller size coils.

The authors recommend that any VLF purchased have the capability of using more than one searchcoil size. Searchcoils larger than eight inches in diameter are available from several manufacturers. Treasure hunters are finding that VLF's when equipped with larger searchcoils such as the ten-and-a-half-inch and even the fourteen inch, make good coin, relic, and small object detectors, plus the scan width is increased considerably. Pinpointing with the larger searchcoils is more difficult at first, but exacting pinpointing techniques can be learned quickly.

The larger searchcoils, fourteen to sixteen inch sizes, are producing great depths when used to hunt for money caches, ore veins, pipes, electrical cables, etc. Of course, these large searchcoils are heavier and more difficult to carry around than smaller ones, but when you need all the depth you can get, the larger searchcoils are required.

Many improvements in controls have come about. Most of the single turn ground controls have been replaced by ten-turn controls. This generally gives the wide latitude so necessary to achieve the desired ground canceling operation, even over heavily mineralized ground.

An improvement that many manufacturers have ignored is the factory setting of the TR discriminate control at the true TR, non-rejection point when the control is rotated completely counterclockwise (toward "zero" discrimination). Should the

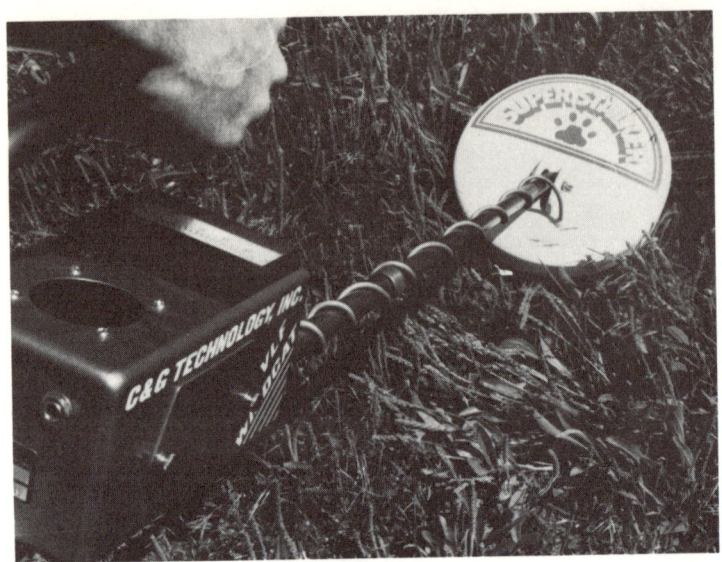

For those who prefer them, there are many smaller, lightweight versions of VLF/TR's available today. Obviously, the smaller the detector, the better that it can be packed and transported. Many coin hunters and other treasure hunters prefer lighter weight detectors. They spend many, many hours each day in the field searching. Every ounce that is added to a detector's weight begins to tell on the operator toward the end of a long day! Good balance, detector stands, adjustable length stems, and submersible searchcoils are a few of the features that manufacturers employ to improve the quality and operating features characteristic of their detectors.

operator of one of the non-factory-calibrated instruments want to conduct ore tests, it might be impossible to do it accurately. If your goal is to purchase a VLF with universal application abilities that include prospecting, study the various detector brochures. Metal/mineral ore sampling, nugget hunting, rejection of mineralized hot rocks, etc., should be discussed sufficiently so that you will know without doubt that these things can be accomplished with the VLF you are thinking of buying.

Push button tuning has been installed on most VLF's. This is certainly a noteworthy method of tuning and greatly facilitates the tuning and operation of the detector. A further improvement on this type of tuning is now available. Mode selection is incorporated into the handle tuning feature. With the slight pressing to one side or the other on a handle-mounted toggle switch, the mode of the detector is changed, electronically, with tuning accomplished at the same time. Press the handle switch to the left for about one second and release; the VLF ground canceling mode is activated and the detector is retuned. Press the handle to the right for about one second and release; the TR discrimination mode is activated and the detec-

With today's modern manufacturing techniques and plastic injection molding capabilities, extremely sleek, well designed control housings can be produced. Not only do present manufacturing techniques lend themselves to good designs, but they also help to speed up and facilitate rapid production of electronic equipment. All manufacturers spend considerable time, effort, and money in continual efforts to update and improve their detector instrumentation.

tor is retuned. The necessity for using both hands to achieve normal operation is eliminated. One-handed operation is achieved, greatly speeding up operation, especially when searching for objects you want to identify, such as coins, rings, nuggets, brass or lead artifacts, etc. This simplified conversion switch is generally described as a "master control switch."

CHAPTER 3

Theory

Basically, what we have discussed so far is the fact that the new VLF detectors are actually Very Low Frequency Transmitter-Receiver instruments. We need to discuss now why they are different from the standard TR. By comparing the two, you will be able to understand the VLF's more easily. This discussion will be divided into two parts. The first part will be a simplistic analysis of the VLF and how it works. The second part will be a much more theoretical and mathematically oriented statement, with actual design data and formulae used by circuit design engineers.

Standard TR detectors are the types which have been around for several decades. Most of them operate within the frequency range of 50,000 Hertz up to 100,000 Hertz. This range is quite a bit higher than the VLF range of 3,000 to 30,000 Hz. The difference between these two types, however, involves more than just the frequency range. (Low frequency circuits described as LF detectors cover a rather wide operating range and the term is rather general. This range operates from 500 to 25,000Hz.)

In order to understand how the VLF types work, it is necessary to take a brief look at how a signal is transmitted into the ground and how a portion of that signal is reflected from the target back to the receiver coil. Almost all types of metal detectors are designed with oscillator circuits. These take direct current (DC) battery power and convert the DC into AC (alternating current) power. The design determines at which frequency the oscillator is to operate.

When an engineer determines an oscillator frequency, he then designs the searchcoil for the best operation at that frequency. The searchcoil is constructed with one or more transmitter coils and one or more receiver coils. The AC oscillator power is fed into the transmitter coil(s). The coil then generates an invisible electromagnetic field which radiates from the coil in

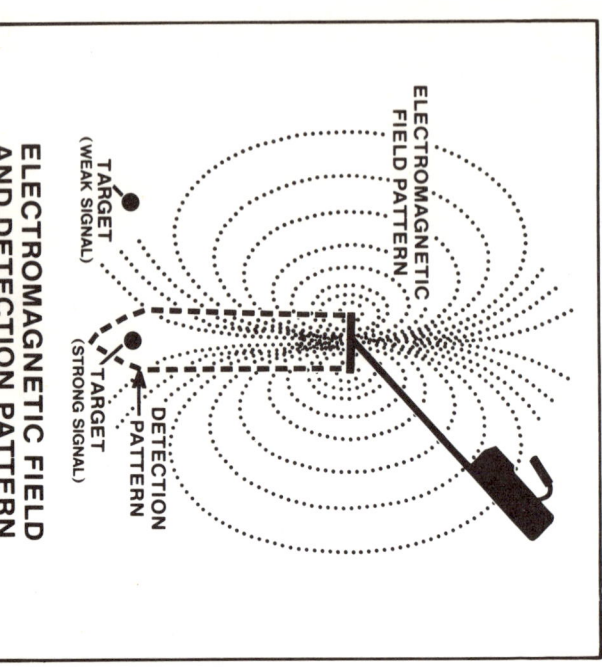

Fig. 2. A representation of the electromagnetic field surrounding a metal detector searchcoil. Note the concentrated field directly above and below the center of the searchcoil which "creates" the **DETECTION PATTERN**. Note how the field strength decreases toward the "bottom" of the **DETECTION PATTERN**.

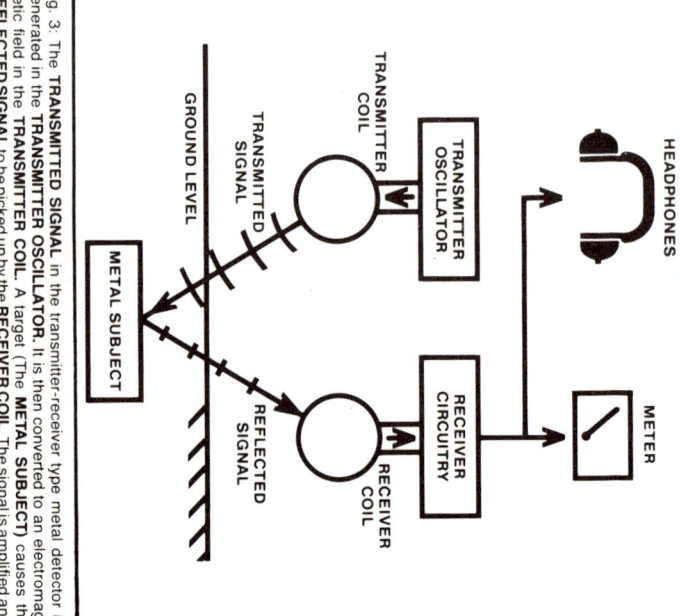

Fig. 3: The **TRANSMITTED SIGNAL** in the transmitter-receiver type metal detector is generated in the **TRANSMITTER OSCILLATOR**. It is then converted to an electromagnetic field in the **TRANSMITTER COIL**. A target (The **METAL SUBJECT**) causes the **REFLECTED SIGNAL** to be picked up by the **RECEIVER COIL**. The signal is amplified and fed to the **HEADPHONES** (speaker) and **METER**.

all directions. If this field could be seen, it would look like a doughnut which encircles the searchcoil.

The downward portion of the field penetrates the ground. If metal is buried, the field also penetrates the surface of the metal. Detection begins at this point. When the field penetrates the surface of the metal, eddy currents are generated on the surface of the target. These very tiny circular AC currents in turn, generate their own electromagnetic field which radiates out in all directions in the same manner as the searchcoil field. A portion of the eddy current field is radiated upward and is received by the searchcoil. The electric circuitry employed then causes audio or visual indication of the metal target.

The return signal is also affected in another way. Whenever iron minerals are present in the ground, the electromagnetic field of the transmitter coil is changed in strength or shape. A portion of this field strength change or variation is picked up by the receiver coil. Thus, when a detector is transmitting and receiving a signal over ground containing iron, TWO signals reach the receiver coil; a signal from the metal target and a signal due to the presence of iron ground minerals. These signals mix in the receiver coil to form a composite signal. In a standard TR detector (not ground canceling), this composite signal is amplified to power a speaker and meter. Thus, it is easy to see why non-ground canceling detectors are so difficult to operate over iron ground mineralization. Raising the searchcoil even a fraction of an inch while scanning over iron mineralization causes "false" signals to be produced simultaneously with the signals from metal targets.

In ground canceling detectors, there are ways to eliminate the "false" signals from ground iron minerals. A phenomenon in electromagnetic signal transmission called phase shift occurs which delays some transmitted and received signals from metal targets. Phase shift has been known and used for many years, especially in radar transmission which was in use during the war years.

This slowed down or delayed signal from metal targets can be separated from the signal caused by iron ground minerals. The signal separation can be accomplished in various ways, but

Recessed, protected control panels are a great step forward in protecting controls and in keeping control knobs from being turned when the detector is used in heavy brush areas.

the basic principle simply stated is that the circuitry is "blinded" to the iron ground signals, but "sees" the metal target signals.

If the question arises as to why standard high frequency detectors cannot be made to cancel iron earth minerals, actually, they can! The depth to which these instruments will detect their sensitivity becomes less as the frequency increases. In the VLF range, the sensitivity is very good. Two main reasons support this fact.

The first is that the eddy current generation is much better. As explained earlier, eddy current generation occurs on the surface of the metal target. The VLF frequencies generate eddy currents more deeply into the surface of the metal. These increased eddy currents cause more electromagnetic field to be re-transmitted to the receiver coil.

The second reason is that the electronic circuits no longer have to contend with the "false" signals caused by iron ground minerals. Much greater circuit amplification can be used which permits the operator to hear weaker signals from deeper metal targets. Research is going on and additional VLF ground canceling detector improvements will surely continue in the next few years.

Robert Podhrasky, Chief Engineer, Garrett Electronics, prepared the following paragraphs for those of you who wish to study the theoretical design aspects of VLF detectors in greater detail. Basic, detailed data that describes the theoretical design and functional aspects of the major electronic building blocks (circuits) utilized in certain VLF/TR metal detectors is given. Each block is pictured or described in the following paragraphs and then the operation of the system is traced so that you might understand more fully how one type of VLF ground canceling detector works.

DEFINITIONS

A. SEARCHCOIL — consists of a transmitter and receiver coil (or coils). The transmitter element is usually a part of the resonant tank of the search oscillator. There are many types of searchcoils and many variations within the types. Some examples are shown. Configurations **a, b, c, and f are** most common.

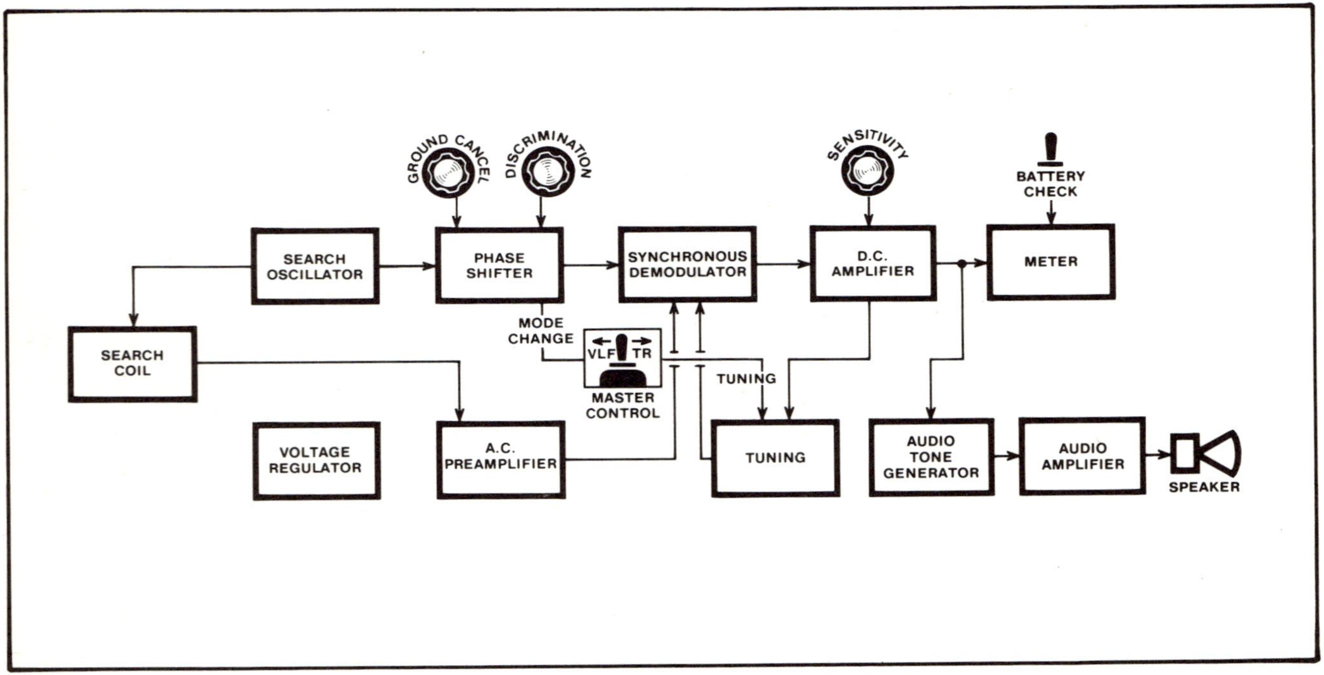

Fig. 4: Block diagram of a typical VLF/TR metal/mineral detector circuit.

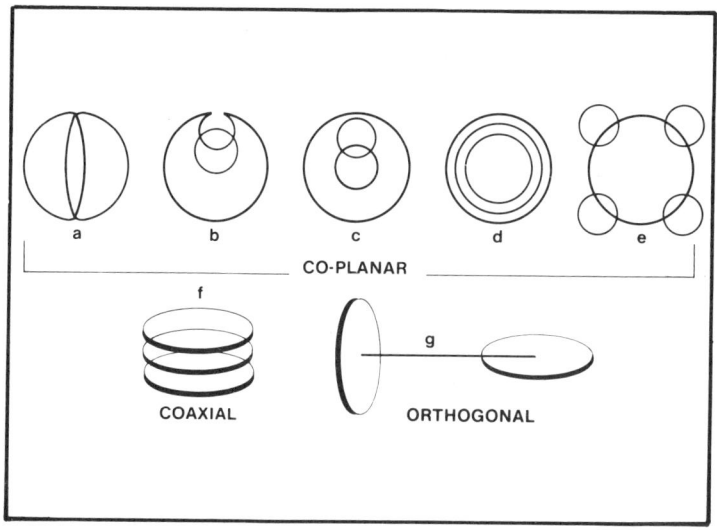

Fig. 5: Illustrated are some examples showing the coil winding shapes used in metal detector searchcoils. **a, b, c,** and **f** are most commonly used in today's metal detectors.

B. SEARCH OSCILLATOR — generally a Hartley or Colpitts oscillator operating in a continuous mode. Some detectors employ an AGC circuit to control the amplitude of the oscillator for better detector stability. The searchcoil is generally a part of the oscillator tank circuit for efficiency reasons. The energy in the transmitter coil is conserved so the equivalent energy in the electromagnetic field is five to ten times the energy supplied by the batteries.

C. ALTERNATING CURRENT (AC) PREAMPLIFIER — a low noise stage employing transistors or integrated circuits. The purpose is to provide voltage gain, as well as band width (frequency) limiting which minimizes noise problems.

D. PHASE SHIFTER — provides the necessary alternations of the phase reference for the demodulator. The ground canceling and discrimination abilities of these detectors depend on the ability to select only the portion of the receiver signal which has a particular phase relationship to the primary field. (This will be discussed in more detail later.)

E. SYNCHRONOUS DEMODULATOR — the circuit that converts the composite AC receiver information into usable DC signals. Operating much the same as a chroma demodulator in a television set operates to extract color information from a composite video signal, this demodulator extracts desired phase related information from the detection signal.

F. DIRECT CURRENT (DC) AMPLIFIER — brings the voltages up to a level required to drive the meter and the audio circuits. Operational amplifiers are used here because the necessary gains of 60 to 80 decibels required circuitry with high open loop gain with low noise characteristics.

G. PUSH BUTTON TUNING — adjusts the circuitry, on command, to a pre-set operating point. Almost perfect stability (leakage currents of no more than a few trillionths of an ampere) is required in order to achieve push button tuning. The technology to achieve that kind of stability was unheard of in the consumer market before the mid-'seventies.

H. METER — used to give visual indications of detection signals; helps indicate the tuning level of the detector; provides a battery check; helps in pinpointing, etc.

I. AUDIO TONE GENERATOR — employs an astable multi-vibrator and a chopper to convert the DC detection signal into an alternating current wave-form of the correct frequency. The output of this circuit is an amplitude modulated square wave, the amplitude of which is proportional to the detection signal.

J. AUDIO AMPLIFIER — provides audio power for the speaker.

K. VOLTAGE REGULATOR — insures the circuitry operates from a constant voltage source. High gains with low power consumption make such circuit elements a necessity rather than a luxury.

GROUND CANCELING/DISCRIMINATION THEORY

Let's trace the events which lead to an indication of a metallic target. The Search Oscillator sets up an alternating electromagnetic field in the transmitter portion of the Searchcoil. The receiver portion of the Searchcoil is positioned so that the net primary energy in this loop approaches zero. The ideal case of zero energy cannot be realized so the push button tuning circuit is employed to offset any residual signals and to set the circuitry to the necessary operating levels.

When a target enters the electromagnetic field and modifies it, a small signal appears at the output of the receiver Searchcoil. This AC voltage is amplified by the AC Preamplifier. The Synchronous Demodulator extracts information from the amplified AC signal, using as a reference the signal from the Search Oscillator which has been modified by the Phase Shifter. The output of the Synchronous Demodulator is a DC voltage. This voltage is then amplified by the DC Amplifier and converted to amplitude modulated AC by the Audio Tone Generator. The power is boosted by the Audio Amplifier before it reaches the speaker.

THEORY OF OPERATION

Thus, we have seen the structure of the workings of one type of VLF circuit. The searchcoil, oscillator, amplifiers, and tone generator operate in a straightforward manner. The key to discrimination and to ground canceling is the ability to extract phase-related information from the receiver wave form. To see what happens when a target enters the electromagnetic field of a searchcoil, consider first a perfectly tuned (mechanically aligned) searchcoil. While transmitter voltage is large, a 10 Vpp. (volts peak to peak) sinusoid, for example, the voltage across the receiver coil is zero (since the coil is perfectly tuned).

When a target enters the primary field, the field is disturbed and the balance in the receiver coil is upset. The receiver voltage will increase as the target approaches the searchcoil.

A short investigation into how and why the electromagnetic field is disturbed will help show the principles for both ground canceling and discrimination. A target may affect the electromagnetic field in two distinct ways. The first to be considered is a result of a redistribution of energy in the electromagnetic field, known as the *reactive* effect. If a conductive target enters the field, eddy currents flow on the surface of the conductor and the field is excluded from the volume of the conductor. The energy in the field is conserved; that is, the total

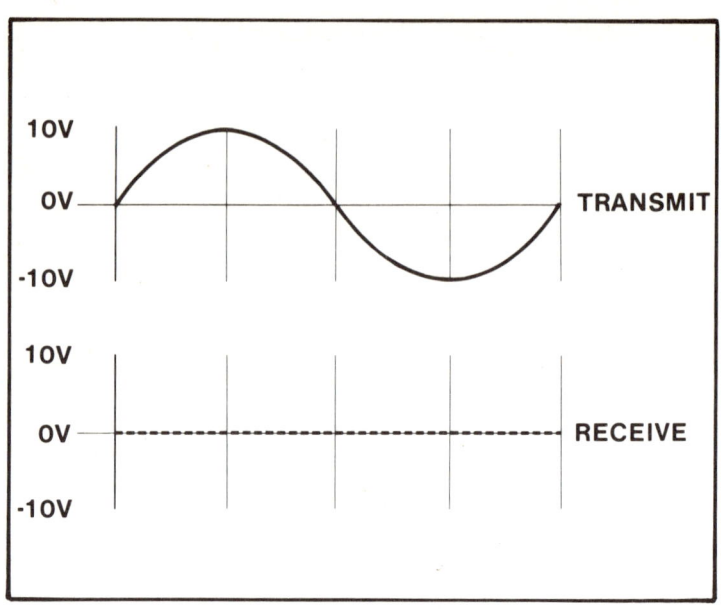

Fig. 6: In a metal detector with a perfectly tuned searchcoil, the amplitude of the receive signal is zero when there is no target disturbance.

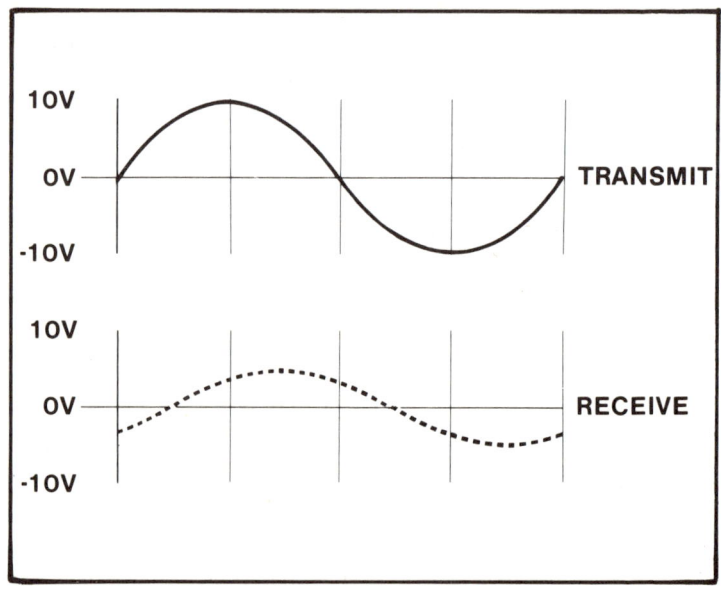

Fig. 7: When a target enters the searchcoil's electromagnetic field, the receive signal is no longer zero.

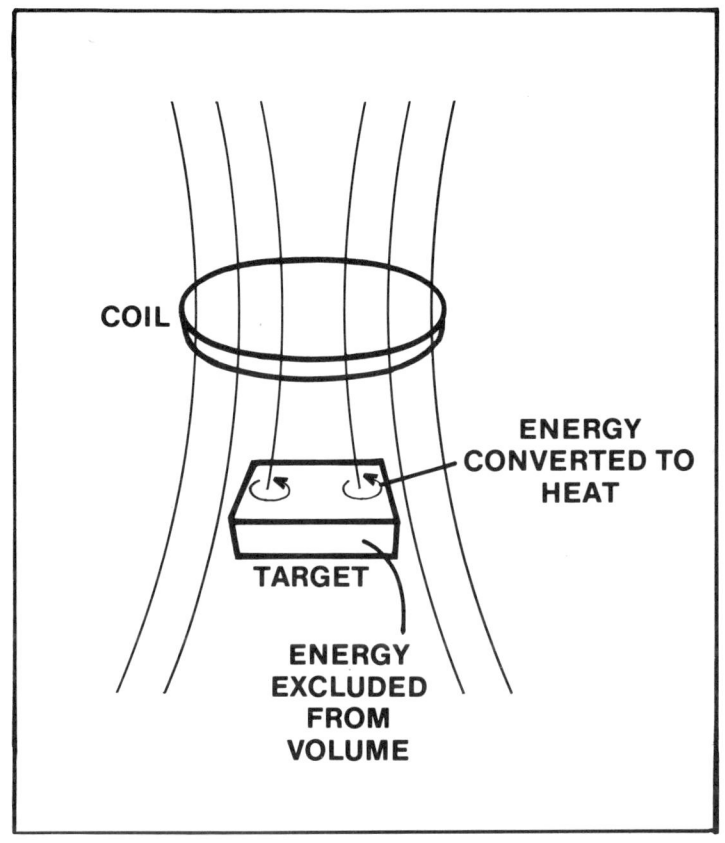

Fig. 8: When a target enters the electromagnetic field, the shape of the field will change and in most cases some of the energy from the field will be converted to heat.

energy of the electromagnetic field remains the same. The energy excluded therefore must occupy some other space. An opposite effect occurs when a non-conductive object with a permeability greater than that of air enters the field. Since the magnetic resistance is lower inside the volume, more of the energy of the field will reside inside this volume. These two effects, although opposite, both result in a change in shape of the electromagnetic field.

The second way is that of an absorption of energy known as the *resistive* effect. The conductive target discussed previously is never a perfect conductor. The surface has a resistance and the eddy currents induced in the surface by the electromagnetic field must flow through the resistance. The result of this current flow is that some of the energy of the electromagnetic field is converted to heat (although miniscule and imperceptible). The voltage change on the receiver coil is, however, measurable and different from a change caused by the reactive effect.

If, as in the case of conductive iron, the energy in the volume is both increased due to high permeability as well as decreased due to a conductive surface, the net result is the function of many factors. The skin depth, or depth in the material in which the eddy currents flow, is a function of conductivity, permeability, and the frequency of the alternating electromagnetic field. These effects become important when choosing the operating frequency of a metal detector.

Perhaps the next question is ... of what use are these reactive and resistive effects? The answer is that they provide the data we need to accomplish ground canceling and discrimination.

Consider ground canceling. We find that ground is basically non-conductive. These soil conductivities range from 10^{-4} mhos/meter for dry ground to 5 mhos/meter for sea water. As a comparison, the conductivity of copper is 6×10^3 mhos/meter. If we have no conductivity, then we have no resistive effect. To achieve ground canceling we need to observe the resistive effect while ignoring the reactive effect.

As to discrimination, in most cases, fortunately, the more desirable targets have a lower electrical resistance. Not only do precious and semiprecious metals have a high conductivity, they are also found in a highly refined state; hence, the result is that their surface has a relatively small electrical resistance. For most undesirable objects, this is not the case. Iron and steel have only moderate conductivities. Although aluminum foil and aluminum foil coated paper have a moderate conductivity, they are often less than one skin depth in thickness and produce a large amount of loss in proportion to the volume displaced. Other objects, such as aluminum ring tabs and scrap metal, generally show up as lossy targets. Unfortunately, nickels and some rings have a fairly high loss factor. Thus, it happens that discrimination is not always 100-percent accurate, although it is, in most cases, a useful tool.

In order for detectors to work properly in high weeds and grass, especially when the vegetation is wet, detector searchcoils must be 100 percent electrostatically shielded. This shielding is often referred to as Faraday shielding, a common designation for electrostatic shielding. Many years ago a Mr. Faraday developed the first electrostatic shield. This is not an RF shield but a DC shield that prevents electrostatic build-up which can be heard through the speaker. You can test the effectiveness of a detector's electrostatic shield by tuning the detector to achieve a very slight threshold of sound, then moving the searchcoil slowly through wet vegetation. You may also pull a handful of long weeds, wet the weeds, and rub them across the coils. If you hear a faint indication or scratching sound, you know that the coil is not effectively shielded and might possibly give you some problems. Most manufacturers, however, are extremely conscious about the way in which they build their detectors and do everything they can to produce sturdy, well built instruments that will give long, rewarding service.

We have seen that almost all targets have loss, some more than others. Reading only the amount of loss is not sufficient because as the object is intercepted by a larger amount of energy, a larger amount of energy is dissipated. For this reason, we must correlate the loss information with some type of size and proximity information. Fortunately, the reactive component contains this information. To discriminate, we compare the resistive and reactive characteristics of a material. By employing the discrimination control we may adjust the metal detector to produce a negative signal for a target with a proportionately high loss factor and produce a positive signal for a target with a lower loss factor. In one of a number of ways we adjust the coefficients A and B to give us the desired result from the first order algebraic equation

$$A(X) - B(R) = C$$

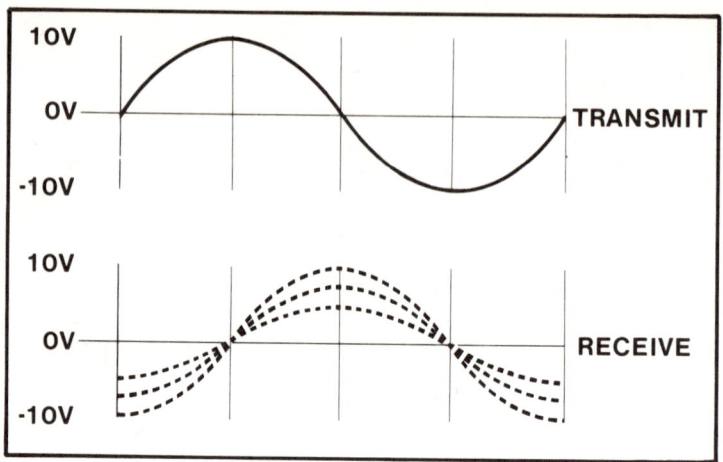

Fig. 9: The quadrature signal produced by the resistive effect of a target increases in amplitude as the target comes closer to the searchcoil.

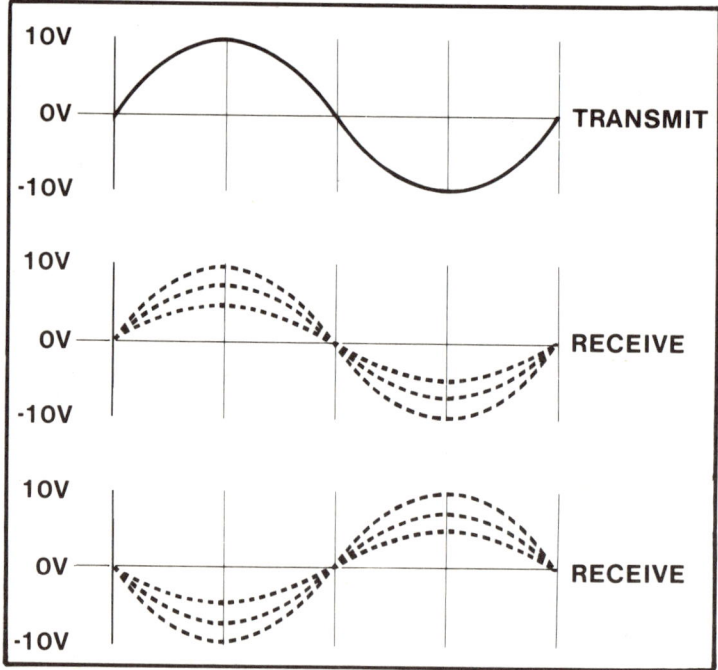

Fig. 10: The in phase signal produced by the reactive effect of a target increases as the target comes closer to the searchcoil. Note the signals are of opposite polarity for energy excluding and energy intensifying objects.

where X is the reactive component and R is the resistive component. When the discrimination control is adjusted for more discrimination, the effect is to decrease the coefficient A and increase the coefficient B. For ground canceling, A = 0 and B = −1. For canceling in conductive ground, such as wet soil or beach sand, A may be adjusted to be slightly positive, thus, the X component offsets the R component of the wet soil.

Now that we have learned that there are distinct responses due to the resistive and reactive components, the next topic concerns how we can extract this information from the wave form. For simplicity we will consider a system with a perfectly tuned searchcoil so in the absence of an object no energy is imparted to the receiver coil. If an object entered the field which resulted only in the resistive effect, a voltage would be induced in the receiver coil which would be in phase in quadrature (a 90° phase shift) with respect to the primary voltage. As the object moved into the area of stronger field, the amplitude would increase but the phase relationship would remain fixed.

On the other hand, if an object entered the field which resulted in only a reactive effect, the voltage in the receiver coil would be in phase with the primary voltage. Although energy excluding and energy intensifying objects cause amplitude excursions in opposite directions, the phase relationship would be the same.

Since most objects produce both in-phase and quadrature components, what we actually see is a composite wave form. An expression for this composition is:
$$A (\sin\theta) + B (\cos\theta) - C \sin(\theta + \emptyset)$$
where $\emptyset$ is a phase displacement. The purpose of the synchronous demodulator is to extract this information with the correct composite phase relationship. Rotating the ground adjust or the discriminate control varies the phase $\emptyset$ to obtain the desired results.

Now, comparing again the VLF type detector with the standard TR type detector, there are two differences. First, the operating frequency of the VLF type detector is much lower. This provides a balance between the resistive and reactive signal produced by targets. Second, the VLF type detector employs circuitry which can extract proper information from the resistive and reactive components of the receive signal. Thus, the VLF (VLF-TR) type detector is a TR with ground canceling and discrimination.

Part 2

Understanding the Operation and Capabilities of the VLF and Other Detector Types

CHAPTER 4

Operating Characteristics of the VLF Metal/Mineral Detector

The Very Low Frequency detectors operating in the ground canceling mode have two faults: they will respond "positive" to out-of-place hot rocks (any rocks containing a sufficiently different magnetic iron content than the matrix to which the detector is tuned) and they are super-sensitive to small pieces of ferrous iron. The hot rock response is troublesome in some areas, but the detector will not positively indicate their presence unless the hot rock is rather close to the searchcoil, a phenomenon known as the near field or antenna effect. If the VLF detector has a TR discrimination mode, this response to highly mineralized rocks is easily overcome. When a detected target is suspected, simply switch the circuit into the TR discriminating mode and pass back over the target. If the sound decreases or dies down completely, this indicates the target is a rock or mineralized hot spot. If the audio remains the same (does not decrease) or perhaps increases, investigate the target. It has some type of metallic (conductive) content.

A word of caution here. Unless you are coin hunting, during this type of metal vs. mineral identification be certain the discrimination control is set at zero. No amount of discrimination is desirable as the target might be a man-made piece of ferrous iron, and any amount of discrimination could cause you to leave a valuable target. Identification of the target as a hot rock should be your only objective, not discrimination. This method of operating is successful inside mines and will successfully identify the hot spot in the matrix as either a conductive ore stringer or a magnetite seam or iron dike. This detection of magnetic and non-magnetic ore veins simultaneously while operating in the VLF non-discriminating mode is considered a great asset in electronic prospecting and was never possible until the introduction of the VLF type detector. After the initial identification, you may then proceed to increase the discrimination control gradually to determine if the target is low conductive pyrite (worthless) or non-ferrous (valuable) conductive ore. We will describe these operating procedures more fully in Part 3.

The tendency of a VLF to be super-sensitive to small pieces of ferrous iron can be worked around; although if the VLF detector does not have the twin-circuit feature, you will be forced to dig most of these small trash targets when coin or treasure hunting.

If the detector is a combination VLF/TR (discriminator), you may adjust the discrimination control prior to starting a coin search. If you fail to identify the target by audio response when operating in the VLF ground canceling mode, simply flip the mode switch to change the circuit into TR discriminating mode and pass back over the target. Identification will be accomplished by the audio sound which will decrease for bad or increase for good. If the area is extremely trashy and the VLF detector has only a manual tuning control, you may experience some difficulty in isolating each target for identification. However, most of the manufacturers use push button or master control tuning which provides instant retuning. These methods permit you to detune the detector quickly while over a target and completely isolate each target, pinpointing it perfectly. During the first few years of commercial development of the ground canceling VLF detectors, some difficulty was experienced in these operations. Now, however, the problems have been almost completely eliminated by the addition of the TR discriminating circuits and instant retuning methods. With sufficient instructions even a young child can easily operate these sophisticated, super-sensitive instruments.

The VLF/TR detectors are well known for their deep seeking capabilities. With the use of average size searchcoils, these detectors will detect coins and small gold nuggets at great depths even *through the most highly mineralized rock or soil.* The availability of large searchcoils for the VLF has made other types of detectors obsolete for deep cache searching. With LARGE coils the VLF detector will go just as deeply as the RF deepseeker type and has the advantage (or disadvantage) of responding to smaller targets. Also, it can utilize different methods of discrimination and is NOT affected by ground mineral content (magnetic iron), as is the two-box RF type. These features give the VLF many great advantages over other types of detectors. The VLF detector can be used to search for both conductive and magnetic ore veins in one operation and will operate perfectly inside mines and caves.

It is possible to conduct mine surveys with electronic detectors and be able to penetrate the mineral matrix. The detection of stringers and missed pockets of high grade ore may eventually make the VLF type detector the most productive of any detector used in the mining industry.

Because of its very low frequency operation and the type of searchcoil construction, some VLF detectors are more or less affected by power lines, electric lights, and other electrical interference. Erratic meter and audio response will result. Inspect carefully before buying. Compare VLF detectors and select one suited to your needs. There are many different types of searchcoils wound in many different configurations, both co-planar and co-axial. Some will pinpoint better than others and some prevent interference from outside electrical power sources. You must take all these factors into consideration when contemplating the purchase of one of these advanced design instruments.

Do not confuse the VLF/TR ground canceling type of detector with TR (induction balance — IB) or LF (low frequency) models popular for single coin hunting. These TR detectors operate at a higher frequency and are generally regarded as shallow detection types that give good penetration on single coins in mineral-free areas.

It is advisable to select an area of search where trash is not abundant while learning to use your super-sensitive VLF/TR type detector. You will thus have the opportunity to become familiar with the detector controls and avoid first day frustration and depression.

Remember that while you may have the latest method of detection available, with the deepest penetration and ease of operation that the most advanced electronic manufacturers can design and build, it is up to you to find spots where coins or relics actually exist. It is foolish to continue to search in an unproductive area. If you find small trash you would also find coins and other valuables if they were there. Some older parks are very littered with trash; some are not. The ones which do not contain much trash but do contain valuable coins, really pay off for the experienced VLF/TR operator. Searching for relics of all types around old cabin sites, battlegrounds, etc., is also a good way to start your first few days of metal detector operation. These types of areas always produce success to some degree and allow you to familiarize yourself with the detector controls and methods of operation.

Any type of discriminating ability is welcome on the VLF type detector if you are to use it for coin hunting. Of course, a true, mineral-free discrimination method by meter or audio would solve the problem. However, it is highly unlikely that this type of discrimination can be developed to 100-percent accuracy because the magnetic iron (Fe_3O_4) in the earth's surface is of the same magnetic iron content as in the common nail. If you were to file a nail into dust, the filings would affect the searchcoil's electromagnetic field the same as does the natural iron oxide in the earth. This causes the circuit to accept some bad targets and also to lose some of the good ones.

Some mineral-free discriminators utilize a time constant circuit in an attempt to reject bad targets. This type of instrument must be operated according to the manufacturer's instructions and involves such an operating method as whipping the detector over each target at a certain speed to determine the content or other unusual method. A few highly experienced operators can produce quite well on coins with the whipping method. However, there is no way to tell what you have missed and, of course, this type of discrimination method is absolutely useless for cache hunting or prospecting. Further, due to the excessive speed of operation required and the room needed to maneuver the searchcoil, the whipping method could not be used in tight quarters, mines, building corners, around fences, metallic objects, etc. The lack of capacity to utilize larger searchcoils effectively could also be considered a major drawback for all around use. Other type detectors may have to be operated at a *slow* scanning speed and this results in unnecessary loss of time. Other types detect or reject objects according to some peculiar detection pattern shape and may not reject bad targets below a few inches depth.

Charles Garrett tests a new type of underwater detector along an island in the Caribbean. The testing was done to perfect the locating of Spanish shipwreck coins and artifacts. Charles is an accomplished diver and performs his own test programs on underwater detectors. For an instrument to be successful it must be designed to work perfectly in salty ocean water.

The major problem, perhaps, with the whipping type VLF discriminators is the extra labor and physical effort required to operate the detector in this unusual manner. Depending on the weight involved, the effort can become quite tiresome. Carefully evaluate the evidence. Is your primary object to locate single coins ONLY? If so, can all the members of your family who will be operating the detector whip the weight for long periods? If not, choose one of the dual mode VLF's that have been proved in the field by so many coin and treasure hunters.

Most manufacturers have solved the problem of discrimination in their VLF models by adding a switch whereby the VLF ground canceling circuit can be changed instantly to a true TR discriminate mode. This mode permits discrimination by measuring the amount of conductivity of the target. The detector then accepts or rejects, according to the amount of conductivity present in the target. Of course, the TR discriminating mode does not operate in the mineral-free mode and care must be taken over highly mineralized ground to use the TR mode only for quick identification of the target. This method is quite fast and easily accomplished. Several VLF/TR's permit the use of larger searchcoils with full discrimination. The method described above is also useful in buildings where ferrous pipes and nails abound.

The VLF/TR detector is, as mentioned, switched into a very low frequency TR mode. This is quite important as the instrument is still operating at the same frequency as the VLF (mineral-free) mode. The TR mode will also penetrate the magnetic iron oxides contained in the ground matrix as easily as will the VLF mode. Extra operating skill is required as the very low frequency TR sees the iron oxide more readily than does a standard high frequency TR and operation over mineralized soil is slightly more difficult. The VLF's TR mode is supersensitive, however, and can identify any target as deeply as the VLF mode will detect it. Its use does require some practice and a steady hand. Actually, since you use the TR discrimination circiut only for momentary identification, if the controls are easily accessible only an instant is required for target identification.

Operation controls on some combination VLF/TR circuits have been greatly simplified. Most employ push button or automatic retuning. Some have advanced so far as to have enclosed completely their main control switches (those that perform many functions) in the detector's handle, somewhat similar to the new citizen's band radio controls that are all mounted in the microphone. Such an arrangement allows the operator to control both circuits and the tuning entirely with one hand, leaving the other free to dig... an especially good arrangement for those who enter competitive treasure meets.

The VLF/TR detectors that discriminate according to the amount of conductivity present in the target may be operated at any speed desired. They can be swept rapidly in either mode, slowly maneuvered into tight corners, or remain stationery, regardless of the mode used. This operational capability permits coin or trash identification, building searching, operation in small mines, and reasonable identification of targets of large mass. The circuits used in the VLF/TR types are generally compact, a construction feature which helps eliminate unnecessary failures in the field. You actually get the performance of two different detectors in one unit, both ground canceling (mineral-free operation) and TR discrimination. The instruments with the smallest number of controls possible and with simplified operating procedures are the easiest to operate and most trouble-free.

If you are considering the purchase of a VLF/TR detector, locate a qualified dealer who can explain all the various types and demonstrate how they work. A salesperson who discredits his competitors and who insists that there is *one* best type detector and that the type he stocks can do everything better than any other type is either unreliable or ill-informed. There can be "best" instruments where quality of construction and field performance are concerned, but each of us has individual needs and desires. If you wish a simple, easy-to-operate detector, use the VLF/TR type and stay with the simplified operation procedure. Turn it on and tune it. Nothing could be more simple. If you wish to perform fantastic feats of target identification, high grade rich ore, find deep coins, operate in trashy areas successfully — then you must learn to master the many different functions your detector will perform. Operation of your detector will then become part of your life style and produce many happy hours and monetary rewards for you.

CHAPTER 5

Comparing the VLF with the Standard TR

For all practical purposes, transmitter-receiver and induction balance detectors are the same, and we will refer to them as "TR." They are very popular for coin hunting and some models are relatively inexpensive, partly due to the fact that certain manufacturers mass produce them for the discount store trade. Low cost construction cannot be expected to result in detectors capable of full reliability and full performance field use.

Their quickness of response is *excellent* and *desirable* in the search for coins, but presents a drawback when stable operation is necessary for searching over uneven ground or rocky areas where mineralization is present. Because of erratic operation resulting when the searchcoil is moved up and down over highly mineralized soil, the TR leaves a lot to be desired for all-purpose treasure hunting and prospecting. The larger the searchcoil, the more erratically and quickly the TR responds to the mineralization in the soil. In these situations the TR cannot be kept tuned at its highest peak of operational efficiency.

Attempts have been made to alleviate the problem caused by mineralized ground with various kinds of tuning and mineralized ground control features, but any type of ground control or ground adjustment feature on a standard TR is nothing more than a gain control or sensitivity adjustment. If a control is used to lessen the effect of mineralized ground, the sensitivity is also lessened. This is, of course, necessary in certain areas and, while it does not solve the problem, decreased sensitivity does make detector operation easier.

An auto-tuning feature puts the detector in a completely automatic mode. The tuning never has to be adjusted; simply turn the instrument on and off. While this seems desirable at first, there is a drawback in that the audio tends to overshoot as the auto-mode keeps trying to adjust the tuning in accordance with the metal or mineral background. Also, in the discriminate mode, when in fully automatic tuning, the detector's audio decreases over a junk target, an erroneous or "false" increase in the audio will be heard. This increase sounds just like a "good" metallic target signal. These "false" signal problems will be discussed in detail in Part 3.

The push button retuning system is a completely manual operation. The detector is tuned in a normal manner while depressing the push button control. The push button is released during normal operation. When the tuning must be adjusted to compensate for ground or temperature changes, reset the tuning instantly to your original point by pressing the push button control. If you wish to operate in the completely automatic mode, simply hold the push button control in a depressed position. However, now operation will become slightly erratic over mineralized ground and some sensitivity may be lost. A few manufacturers have installed an automatic mode switch, as discussed in the previous chapter, thus eliminating the necessity of holding the button down.

Many TR detectors have a DISCRIMINATING mode in addition to the NORMAL mode of operation. It permits the detector to be used in the normal mode for coin hunting. All metallic targets are acceptable in the normal mode. Generally, a switch or adjustment control is provided to permit energizing the discriminating circuit. In discrimination, many junk or unwanted trash items (tinfoil, bottlecaps, small ferrous iron pieces, etc.) will be rejected. Many detector models may be adjusted to reject aluminum pulltabs. However, these pulltabs have a high conductivity factor and the increased amount of rejection reduces sensitivity and nickels and small rings may be rejected also. Whether the loss in sensitivity necessary to accomplish pulltab discrimination is desirable depends on the search area and the operator's wishes.

In order to achieve acceptable discrimination, the searchcoil configuration sometimes has to be changed slightly from the standard TR design. Some TR detectors featuring discrimination are more adversely affected by mineralized ground than are non-discriminating types. All the different points should be evaluated before purchasing a TR detector. The twin circuit or dual-purpose unit is more practical for the average hobbyist who has only one detector, but the person who enters speed contests or treasure meets and has no need for discrimination might desire the normal TR detector with a 2D co-planar type searchcoil because of its easier operation over mineralized ground.

Large searchcoils are difficult to use on the standard TR type detector because the solid construction presents a weight problem. Thus, most manufacturers do not promote their large TR coils, preferring to sell the TR detector for the purpose for which it was designed... coin hunting.

Though many caches have been located and many large natural gold nuggets may have been recovered using TR's, the fact remains that because of the quick response operating characteristic, the TR searchcoil does not respond correctly and is erratic when used in highly mineralized areas. The TR is completely inhibited in marshy areas and inside mines or caves that contain wet, mineralized soil and rock which disturb the electromagnetic field produced by the searchcoil. Place a coin an inch deep in a rain-soaked, mineralized area and notice the reduction in the detection signal. You might not be able to find the coin at all!

If you own a TR type detector, perhaps these words about the design characteristics will help you to understand your detector. If you desire to purchase a TR, we suggest you purchase a well-known brand. Unless cost is a factor, the purchase of a standard TR is not recommended because many manufacturers combine a TR discriminating circuit with their VLF type detectors. The VLF/TR type contains the ground cancellation feature, plus a very low frequency TR discriminator circuit.

CHAPTER 6

Comparing the VLF with the PI Unit

The pulse induction (PI) detector operates on a much different principle than do the more common beat frequency oscillator (BFO) or transmitter-receiver (TR) types. It "pulses" a radio frequency (RF) signal into the earth, almost completely *ignoring* iron mineralization (negative) and salt-saturated (positive) beaches. Due to its nature, it ignores small pieces of troublesome tinfoil. It does not have to be adjusted or balanced to the mineralization present in the soil as the VLF types do. It is stable, not affected by weeds or grass, has the same approximate sensitivity as the newly developed ground canceling VLF types. It is lightweight and, very possibly, the easiest detector in the world to operate. All functions are normally controlled with only one to three knobs. At the present, all PI's are imported from England.

The response of a PI to a target is a bell-like ringing and the instrument is usually operated with the aid of earphones. Highly trained professionals in many countries use the PI detector in security type situations. Battery drain is not excessive, considering the sensitivity produced. The PI is manufactured in both completely automated versions and manually tuned models. Various sized searchcoils are available.

Careful attention should be taken here to distinguish the difference between the words "ignore" and "reject." There is a decided difference. To "ignore" a target, good or bad, means the detector response neither increases or decreases. To "reject" a target means the audio or meter response decreases when the searchcoil comes within a target's detectable range.

The PI can generally be moved, without retuning, from a section of sand containing iron mineralization (negative) to below the tide line where the salt (positive when wet) will cause a positive response on most other detector types. The PI is the only type detector capable of detecting a man-made ferrous target (tin can or metal box) in an ore dump that contains both negative (magnetic iron oxide) and positive (conductive) pyrites and low grade ore. The instrument simply ignores both the pyrites and low grade ore.

After reading the above paragraphs, your first reaction is very probably that you have found your first "perfect detector." Some future day this may very well be, but until knowledgeable electronic engineers correct the faults of the PI, you will be better equipped with the VLF/TR types both for sensitivity and ground canceling (mineral-free) operation.

Let's consider the PI's capability of ignoring magnetic iron mineralization. This means, for one thing, that the PI cannot identify metal vs. mineral. Perhaps your search area contains hot rocks, rocks that contain considerably different magnetic iron content than the matrix (earth's soil composition) to which your detector is tuned. The PI responds positive to these out-of-place hot rocks, just as the VLF/TR types do. The drawback of the PI type is that it cannot be switched into a true TR (non-discriminating) mode to identify the target correctly as either metal or just a highly mineralized (magnetic) rock. You are forced to dig it, for if you don't you may perhaps leave a valuable treasure. This situation will occur frequently if you are searching for nuggets in old dredge tailings, placer deposits, and old mines. It can also happen right in town when coin hunting. Your only alternative is to carry a BFO type detector to identify each detected target correctly or to use a VLF type which incorporates a TR discriminating circuit. You will encounter the same problem with a PI when cache or treasure hunting. Unless you are searching an area completely devoid of mineralized rocks, you will spend your entire day trying to guess the target — whether hot rock or metal.

Stability is a good feature found in the PI, but many detector manufacturers produce VLF/TR types that are just as stable. A PI is unaffected by grass or weeds, but many manufacturers use Faraday shields on their searchcoils, again leaving the PI no edge. Lightweight? A number of VLF/TR detectors are produced in very lightweight models. Simple, easy operation? The VLF/TR types have the advantage of twin-circuit operation and some manufacturers have eliminated unnecessary control knobs, manufacturing their more sophisticated instruments with TWO circuits in ONE detector with one-handed operation.

Of course, cost should not be considered if one is desirous of getting the best, most practical detector, but paying exorbitant prices and not getting the best is not good American knowhow. The response of the English PI, a ringing, bell-like tone, is rather difficult to use to pinpoint and presents a problem where small trash abounds. The VLF/TR types generally have push button tuning. Detuning of the instrument to allow for fine pinpointing is accomplished quickly and easily. The average PI operator finds it difficult to identify targets, while VLF/TR operators have the advantage of the TR discriminating circuit that rejects bad targets, according to the conductivity of the metal. Battery drain is about the same as on some of the higher quality VLF types, although it can be said that there are a few VLF's which have much more battery drain than others.

There is no great accomplishment in designing PI's in both automatic and hand tuned models; most large manufacturers produce VLF's with a choice of automatic, semi-automatic, push button, or "master" control switch tuning controls. You have the advantage of choosing a variety of methods of tuning.

The PI type will not respond to iron pyrites (conductive) and some conductive ores that are extremely valuable so it cannot be used effectively for prospecting. It also ignores small gold nuggets. As you can see, the minus side has almost completely canceled the plus side, leaving only the beachcombing operation as the area in which the PI will perform effectively. Even here, however, it cannot lay claim to perfection. It is difficult to pinpoint a target and the PI does NOT have a discriminating circuit as do most VLF/TR types.

Perhaps someday someone will produce a lightweight version of the PI type with full discrimination. Until that day, however, the very capable and highly versatile VLF/TR type will continue to dominate the industry in sales both to coin and treasure hunters.

CHAPTER 7

Comparing the VLF with the RF Two-Box

Until the development of the VLF type detector which utilized the large co-planar searchcoils, the two-box radio frequency (RF) detector, actually a transmitter-receiver type, was rated as the deepest seeking instrument on the market. As with any detector, the two-box type has advantages and disadvantages.

This type detector is used primarily to locate buried pipes, cables, magnetic or non-magnetic ore bodies, and other large metal objects. Special two-box RF models are manufactured specifically to locate pipes and cables. These pipe finders incorporate a wire that is connected between the transmitter box and the pipe or cable. The receiver box is carried along the ground above the buried portion of the pipe or cable, and the detector picks up the signal being transmitted through the pipe or cable. For this purpose the two-box RF has no equal. This kind of commercial pipe locator is normally sold by a cable or pipe supply company. Some metal detector dealers, however, still stock the regular two-box deepseeker RF without the pipe and cable hookup. This model has been used for years in attempts to locate magnetic and non-magnetic ore bodies.

The complete inability of any type transmitter-receiver to identify metallic ore correctly in the presence of high mineralization (Fe_3O_4) has resulted in many failures and disappointments for miners. This two-box metal detector is not an aid for prospectors. Many failures attributed to the two-box RF detector actually should have been attributed to poor judgment in selecting the TYPE of detector required for the job. When the two-box detector is employed for ore body detection, it must be remembered that it may respond either to metal or mineral. Any rock containing a high degree of mineralization may respond the same as metal and, as a result, many empty holes have been dug in areas of high mineralization.

George Mroczkowski, well-known treasure hunter, assists in a fantastic search for two lost Mexican Army rifles. In the top photo, the owner of the rifle is jubilant as he pulls his gun from the mud and water. Note the two-box RF detectors and the deepseeking VLF/TR detector equipped with the large searchcoil. Note also the probes the Mexicans used in their frantic search to find the missing rifles. George's book, *Professional Treasure Hunter*, describes this successful search and many others.

Almost any underground wet spot can cause a target indication even when the detector is tuned in the metal mode of operation. Hot rocks (containing Fe_3O_4) mixed in with other less mineralized rocks will almost always result in signals which cause useless digging. Many hunters have abandoned searches due to a large number of false signals and poor depth penetration in mineralized soil.

The two-box will usually not detect small metallic objects. However, an experienced operator might be able to detect an object as small as a silver dollar, especially in a non-mineralized area. Despite the problems, the two-box RF has been used for years by many treasure hunters because, until recently, if you wished a deepseeker you were practically forced to use this type detector.

The newly developed VLF detectors with the horizontal loop configuration (co-planar searchcoil) still produce the same false responses on hot or highly mineralized out-of-place rocks. However, VLF/TR dual mode types are capable of identifying these hot rocks. VLF/TR types also penetrate mineralized soil, and there is, in some brands, a choice of searchcoil sizes. The VLF/TR types respond to small targets and will double as coin or nugget hunters. Thus, the VLF/TR types have many advantages over the two-box RF models. Even in spite of the fact that VLF's are not designed as pipe locators, they do as good a job, or better, locating buried pipes as do the two-box RF's.

We cannot overstress that the individual should always carefully consider what his major interest will be and then choose the type of detector that performs best in that particular field. The two-box RF is excellent for pipe and cable location or for location of large targets in non-mineralized soil. If you are considering it for any other phase of metal detecting, make a full investigation of the facts. A test of both types, the two-box RF and the VLF/TR, will quickly confirm the tremendous advantage offered by the horizontal loop style VLF/TR.

CHAPTER 8

Comparing the VLF with the PRG

There is no question but that the phase readout gradiometer (PRG) is the best among discriminating detectors. It uses a gradiometer sensor and operates about the same as does a TR type. Its sensitivity is about equal to that of the VLF and PI types. It is not affected by salt water and it is unbeatable on a beach where there is no magnetic black sand.

There is, however, one major drawback to the PRG. If the soil contains moderate or high amounts of magnetic iron — a very large percentage of the world's soil does — the PRG is almost impossible to operate.

Another drawback may be its cost. Cost should not be considered when choosing a quality instrument, but the highest priced detector may not be the best detector for you.

The PRG is rather heavy, but it is stable, sensitive, and gives almost perfect readouts on targets in non-mineralized search areas. If you treasure hunt in areas free of iron mineralization and can afford the price, you should test this instrument. The PRG is designed to operate in salt water zones and over non-mineralized ground with excellent target identification.

As with any type of equipment, consider and evaluate all evidence before purchase. If you have use for these specially constructed instruments, they are fantastic; if not, select your detector from types with more versatility. Additional technical information may be obtained from the manufacturer's brochure. Charles Garrett's book, *Successful Coin Hunting*, is a good source of information on detector types and usage.

CHAPTER 9

Comparing the VLF with the BFO

In years past the beat frequency oscillator (BFO) type detector was frequently referred to as the only all purpose detector type available. Of course, this did not mean that the BFO would perform ALL of the tasks assigned to it as well as might another type. It meant only that the BFO excelled in some fields of operation and would also perform with reasonable efficiency in all of the others. Many detector types will not do this. The versatility and ease of operation under most field conditions will probably continue to keep the BFO as a favorite of a few hardy souls for years to come. However, it must be realized that the recent development of the new super-sensitive VLF/TR detectors has almost completely eliminated any need for the old faithful beat frequency oscillator detector which has considerably less sensitivity and is adversely affected by iron mineralization.

A BFO may incorporate a discriminating circuit as an aid to rejecting iron, bottlecaps, and other unwanted items when searching for valuable coins. The discriminating mode of operation has no value, however, when used in metal vs. mineral identification. In fact, the meter will erroneously indicate a conductive ore specimen to be bad. The same result will occur on a small natural gold nugget. For prospecting or metal vs. mineral identification, the true, normal mode of the BFO must be used. It then will respond correctly and identify an ore sample as having either a predominant amount of mineral (Fe_3O_4) or metal (any conductive substance). The BFO also has a fixed center of tuning and wide dynamic operating range. The VLF has no "fixed" center of tuning unless the manufacturer has factory pre-set the ten-turn ground control so that its exact center is set to the "center" of tuning. The electromagnetic field pattern produced by the BFO searchcoil is absolutely uniform.

In the absence of mineralization and at shallow operating depths, the BFO discriminating circuits produce good results without false meter indications.

BFO types do not detect as deeply in mineralized ground as do the ground canceling VLF and PI types. However, the BFO does not respond to tiny pieces of man-made ferrous iron. The VLF and PI types are "all metal" detector types and are especially sensitive to all small targets, particularly iron. Of course, if you don't have super-sensitivity and super-depth you are bound to miss many valuable targets. The BFO does not have the sharp audible "whap" response of the standard TR type on small coins. It may, however, be operated on more uneven terrain in mineralized zones with more success and ease than can a standard TR coin hunting type detector.

Though treasure hunting is generally considered to consist of searching for large chests of money or precious metal caches, this is not always the case. There are many small can or jar caches that may contain many thousands of dollars. For this reason you need to employ a detector that can make use of the large searchcoils. For a cache thought to be the size of a tobacco can or bigger, the large BFO coils will always be the best size to use.

Some BFO types make use of independently operated dual coils. This arrangement is a definite advantage since you can switch from a larger to a smaller coil to determine if the target lies near the surface and approximately how large and how deep it may be. Different size searchcoils, from the small three-inch to the large deepseekers, may also be employed on the BFO with a discriminating circuit.

As surely as the automobile retired the horse and buggy, more modern instruments retire the older, less productive types. The BFO type detector has achieved a remarkable niche in treasure hunting history. It will do everything, anywhere, anytime, with some degree of success. Nevertheless, due to the recent development of the VLF's that are free from the effects of iron mineralization, we hesitate to recommend the purchase of a BFO type.

If you already own a BFO or are convinced you need one, perhaps this information will assist you in its operation. Be certain to choose a BFO type capable of accommodating the larger searchcoils. Select one with a discriminating circuit incorporated with normal BFO operation and "zero" drift. These requirements generally eliminate the BFO types currently sold on the mass market. Those instruments are little more than toys, generally proving disappointing to the buyer, and the treasure hunting hobby loses a new member! Buy from the larger manufacturers with a reputation for quality instrumentation.

CHAPTER 10

Comparing the VLF with Discriminators

Discriminators (Rejectors, Analyzers, etc.) are equipped with a conductivity "readout" indicator, either meter or audio or both. This means a control may be set (some are factory pre-set) to discriminate: that is, to "reject" metallic targets of low conductivity or to "accept" targets of a higher conductivity. Generally, ferrous targets (iron) have the lowest conductivity. Such targets include nails, bottlecaps, tinfoil, and other junk often encountered while coin hunting. Non-ferrous targets with a higher conductivity factor (such as coins, aluminum, brass, etc.) will be accepted as "good."

NOTE: Remember, when we discuss "reject," we mean the audio or meter "decreases." "Accept" means the audio or meter "increases." Do not confuse this method of discrimination with the method known as "ignoring" the target. To ignore means neither to decrease nor increase. The "ignore" system does not permit the full range of adjustment necessary to change from metal vs. mineral identification to rejection of large mass targets. The "accept" or "reject" system permits a full range of discrimination. In a detector designed primarily for coin hunting, the "ignore" system would be the best.

You should be aware of the possibility that when your detector is adjusted to reject aluminum pulltabs, nickels (which have poor conductivity) and some small rings may be rejected. Their small metallic mass effect on the detector's electromagnetic field, coupled with their small amount of impure alloys will cause rejection. The discriminating circuit will accept as good a .999-fine gold or silver bar and small rings made from sheet gold or silver. Small rings of approximately fourteen to twenty carat gold or platinum may be rejected. A discriminator will often indicate a small nugget as bad. The rejection is caused by the nugget's rounded surface. Larger nuggets will probably be accepted as good, depending on their shape.

A high quality discriminating circuit that has many practical advantages might reject some small target that another discriminating circuit might accept; yet, the latter detector might be of little practical use in the field.

A discriminating detector with a control permitting "rejection" adjustment is more versatile than one with factory-set discrimination. The latter cannot be adjusted to permit rejection of tin cans and other large metallic targets when you might wish to detect a known object such as a large silver bar, large aluminum mass, brass foot valve, etc.

The height of operation of the searchcoil over highly mineralized ground can be very critical. Discriminating circuits, regardless of the type, should be operated at a constant height over mineralized soil, especially when attempting to identify the target. The ground magnetic content affects them all. Any erratic up-and-down motion by the operator causes the audio and meter indication to change when in the discriminating mode. Though the PI basically ignores the magnetic content of the soil, if it is brought into close contact it may respond either positive or negative, depending on the composition of the soil. Thus, even a circuit that would ignore the magnetic content should be operated at a constant height when discriminating. Ground canceling circuits that use the feed back or time constant principle are also somewhat affected by magnetic soil content, regardless of claims of "mineral free" discrimination. The operator is expected to "whip" or control the searchcoil in such a manner as to cause the detector to respond only to the metallic content of the target. Unusual operating procedures have their problems and are generally a handicap for any application other than coin hunting. The whipping technique may cause some targets to be "lost" or the operator may find it impossible to reject certain types of trash, like bottlecaps. The method can also become very tiring, depending on the size and weight of the detector and the strength of the operator. VLF circuits designed to discriminate in a mineral-free manner have their faults. They sometimes reject good targets, depending upon soil conditions and the depth of the targets. It is better to dig a few bad targets than to miss valuable ones.

Detectors designated as discriminators generally are equipped with only one searchcoil with no provision for use of

Charles Garrett directs the loading of a pack train used by the authors on one of their prospecting trips. The location, deep in Old Mexico, is one of the richest mining districts in the world. Burro trains must be used to pack in the gear and to bring out silver ore. VLF/TR type detectors were used to search some of the old Spanish mines and hundreds of pounds of ore and silver caches were discovered.

larger deepseeking coils. This is a drawback when attempts are made to utilize the instruments for cache and relic searching. As we stated earlier, discriminators cannot be used for metal vs. mineral ore sample identification. The factory-set discriminator is usually designed to operate with only one searchcoil.

Almost any type of detector may be designed with a discriminating circuit, but the ones offering both normal and discriminating modes of operation are the best buy. Keep in mind, however, that discriminators are not 100-percent-accurate since many factors enter into their operation . . . the mineral content of the soil, the conductivity factor of the target, and the expertise and experience of the operator.

The BFO and TR types (including the VLF types) react differently when in the discriminating mode. The TR generally has an adjustment whereby the amount of discrimination may be increased or decreased. The typical method for setting this control is to set it to reject a bottlecap when the cap is about one inch away from the bottom of the searchcoil. If you move the bottlecap closer the detector will accept it as "good." While discrimination may be adjusted any way you wish, if you set the control to reject the bottlecap at a distance much more closely than one inch, you may decrease the sensitivity or cause the detector to reject "wanted" targets. Regardless of the discrimination control adjustment you make, however, the TR will still indicate the bottlecap as being "good" if it is brought into very close contact with the searchcoil. This is an operating characteristic that detectors with co-planar searchcoils have not yet overcome. BFO's do not have this fault, but TR discriminators produce better results than BFO's on deeper, coin-sized targets because the TR's audio circuit is easier to interpret. BFO's give excellent results in trashy areas, but at lesser depths than TR's. Many traveling treasure hunters simply give up on roadside parks and stopping places because of the great amount of trash. However, many good finds continue to be made by those who understand the proper methods of searching. Rings and jewelry are quite often found. People losing them do not recall where the items were lost or do not have time to return to search. (They probably do not have a metal detector, either.)

At roadside picnic grounds there are usually many pulltabs and bottlecaps lying on the surface or very shallowly buried. They will produce positive signals. To insure that the detected object is worthless and not to be dug, the TR discriminator must be raised above the ground approximately one inch and passed back over the spot. The detector will have to be retuned each time. This procedure is greatly aided by the push-button control for instant retuning. You simply reset the tuning, scan over the spot, then reset the tuning after you lower the coil back to the ground for deeper searching. Patience and slow searching procedures are required for surface or near-surface targets when a co-planar searchcoil is used. Co-axial TR searchcoils do not have this fault. Either type coil will produce satisfactory results in discriminating detectors, but the operator must decide which he or she wishes to use.

The VLF/TR has ground canceling capabilities (VLF mode), small and large coil discrimination (TR mode), and produces excellent depth on coins and large targets in almost any type of adverse soil conditions. VLF mode circuitry, however, cannot be balanced or adjusted to eliminate the effect of wet ocean beach sand. The VLF's do not (in the VLF mode) reject the earth's iron oxides but are adjusted to ignore them. This can be called a form of discrimination, but, however, the VLF's do not discriminate between the different types of conductive metal while ignoring the magnetic oxides. In order for the VLF type to discriminate among the different types of metal effectively, they are switched into the TR mode of operation. The low frequency electromagnetic field penetrates the iron oxides with little, if any, loss in depth. When switched into the TR discriminating mode, they are still operating in the *very low frequency* mode and still have the *same penetrating power*. This is very advantageous for a discriminator and produces great depth when correct identification of a small target is desired.

Some manufacturers try to eliminate every unnecessary control and some incorporate most of the main controls into the handle. This kind of fingertip control permits one-handed operation and allows instant mode change and retuning. Many manufacturers also incorporate a switch or control to allow the detector to be operated in the true automatic mode. The circuit then adjusts itself automatically to the earth's mineralization effects. However, as previously discussed, over mineralized and uneven rocky ground the audio tends to overshoot while attempting to keep in harmony with the effects of mineralization. However, a metallic target can still be identified as a more sharp, pronounced response. It is possible, nonetheless, to operate the very low frequency TR types in the full automatic mode with full discrimination. Many prefer this automatic mode of operation when searching the beaches.

Many successful operators switch the VLF/TR into the TR discriminate mode, and set the discrimination control to reject pulltabs and large cans. They decrease the sensitivity control if the soil is highly mineralized and operate the searchcoil from two to four inches above the ground, listening for the increase (good response). Actually, this is making the VLF type perform the same as a BFO, but the VLF will produce better results with this method than will the BFO. Remember, the TR discriminating portion of the circuit is operating in the VERY LOW FREQUENCY range and this permits large cans to be rejected while small coin targets are still indicated.

The correct positioning of a detector's handle is very important. It should be far enough forward to achieve good balance. The angle of the handle also should be tilted slightly upward. This upward tilt prevents mechanical overswing of the detector as it is moved from side to side. White searchcoils are found on most detectors. Though the use of white is not 100 percent necessary, white searchcoils do assist in preventing or lessening drift because the white color helps reflect the heat rays of the sun, minimizing searchcoil heat expansion which can cause drift. Most detectors are built ruggedly and will last the owner many, many years, provided reasonable care is given to the instruments.

In most models, the VLF/TR types can discriminate at VLF depths. If the controls that change the mode of operation are easily accessible, you can operate in the VLF (ground canceling) mode, locate a likely target, instantly switch into the discriminate mode, identify it, return to the VLF mode, and continue searching. Practice will make this method quick and easy, without the loss of any of the depth produced by the mineral-free circuit.

It will be of great help to you if you will study the difference between mineral-free discrimination and very low frequency TR discrimination and then have a manufacturer or dealer explain that difference thoroughly and honestly. You will not come to erroneous conclusions nor expect your particular detector type to do jobs under conditions for which it was not designed. Quality and performance are all-important in selecting any detector. It is absolutely mandatory to select a detector according to the type of work it will be called upon to do.

Part 3

Using the VLF and Other Detector Types in All Phases of Treasure Hunting

How to use the most popular types of detectors in all phases of treasure hunting, with special emphasis given to VLF/TR metal/mineral detector usage.

CHAPTER 11

Searching for Money Caches

Treasure hunting covers all phases of detector usage. In this chapter we will discuss only cache hunting or, rather, the searching for buried money treasures. We will not discuss hunting for coins or small objects. Detectors are pushed to the utmost in depth penetration for the larger targets. Often, the mineral content of the soil determines the response and sometimes the size of the cache may determine the outcome of the search. Many beginners have hit the treasure trail in a desperate attempt to make it pay as some of the full-time treasure hunters have. All have quickly discovered that it takes more than just a detector and ambition.

Many factors enter into the successful recovery of deeply buried caches; as a consequence, few people have achieved enough success to become experienced. A few factors are: the location of the suspected treasure site; the amount of vegetation present; the amount of adverse ground mineral content; the condition of the ground surface; how small or large the cache is reputed to be (a quantity generally overestimated); how deep the cache is buried; whether the ground has been built up over the cache; whether the detector can penetrate the type of soil or rock that is present; whether you have the correct size searchcoil necessary to obtain all possible depth but still small enough to respond to the estimated target size; etc. Do not complete your research and arrive at the suspected location before considering these matters. Since we know of no guide book or manual on the market that explains all these various factors, we will attempt to touch on every aspect of cache hunting, with every type of detector, with the hope of explaining why a past venture failed and to help you achieve success when you arrive at your next location.

The correct equipment must be used. A major failure of beginners is that they expect to find deeply buried targets with a small searchcoil or with the same detector and coil that was so successful in the search for shallow relics and small coins. To be successful, you must use large searchcoils for the recovery of larger, more deeply buried treasures. Regardless of overenthusiastic advertising and tall tales told by inexperienced treasure hunters, you cannot change electronic facts. If an average size searchcoil, say eight inches, produces an electromagnetic field reaching three feet, this does not mean it is capable of detecting an average size chest or kettle at this depth. It means only that you would get a faint indication on a large metallic target, such as a car body.

One factor that some treasures hunters cannot seem to grasp is that the length of time the target has been buried also enters into consideration. If you bury metal, the depth expectation is reduced, depending on the amount of mineral present in the soil and type of detector used. Bury a quart jar full of silver or other metal from ten to twenty inches deep, a large kettle or Dutch oven from eighteen to thirty-six inches deep. Use the best detectors available with average size searchcoils and you will understand why most treasures are simply walked over and *missed* by hard-working treasure hunters who placed their faith in small or medium size searchcoils. Even the best deep-seeking detectors utilizing large searchcoils will experience some difficulty in this test but, as years pass, the targets can be detected more easily.

Most caches are as large as a small tobacco can and may extend to any size, such as a chest, kettle, or even bigger object. This fact allows a wider choice as to type of detector used, but you will still need to employ the large searchcoils as insurance in gaining greater depth. In the following paragraphs we discuss types and established guidelines for successful recovery of deep targets.

No slight is intended to any manufacturer who does not produce detectors capable of utilizing large searchcoils. Many manufacturers produce quality lines solely for the enjoyment of coin hunters. There are some low-quality instruments capable of using larger searchcoils, but they may be unstable and so poorly designed that even the large coils are worthless in the field. Professionals who make their living in the treasure hunting business always demand the best quality instruments and usually own two or more detectors capable of good field performance. After all, money missed is money wasted, and the deep ones you might miss would pay for many good instruments.

CACHE HUNTING WITH THE BFO

The BFO you choose MUST be absolutely 100-percent stable to achieve success, especially when using large searchcoils. The coils must be Faraday-shielded to prevent grass and weed interference. The BFO detector should be sensitive, but not at the expense of losing stability. Some units have the gain advanced so high to impress customers with air tests that the units are unstable in the field and it is impossible to set the tuning speed at a moderate or slow motorboating sound. The slower tuning speed is necessary because you can hear the increase in beats better when you must move slowly because of limited maneuverability with larger searchcoils.

Twelve-inch diameter searchcoils will operate best at approximately four inches above the ground, an operating height which frees the coil from most ground effects and obstructions. Tune at a slow or moderate motorboating sound (twenty to thirty beats per second); scan at the rate of approximately three feet per second. You may decide to change the scanning speed after becoming familiar with the faint increase in beats when over a metallic target. The twelve-inch coil should detect targets as small as a fifty-cent-piece. It will have a maximum depth of approximately four to six feet on large targets (in mineral-free soil). It is the best size to use for suspected caches similar in size to a small tobacco can.

Twelve-inch coils have limited use, however, because larger sizes will detect targets as small as a silver dollar *and* have the advantage of going much deeper on larger caches. Twelve-by-twenty-four, eighteen, twenty, twenty-two-inch, etc. . . . these large coils are very popular for cache hunting. They are small enough to maneuver, yet large enough to get all depth the BFO is capable of producing. If the cache is thought to be deep, then you should employ a VLF type with a large searchcoil.

Many different BFO configurations are available. The extremely large coils designed in round patterns are difficult to sweep widely enough to allow the deep target to respond. Elongated coils, such as 12-by-24-inch, weigh less and produce better target signals because the coil is narrow, though still long enough for fast ground coverage. This size and type of BFO coil is best for caches and it is produced by most BFO manufacturers. That is not to say the elongated coil will go deeper or even as deeply as the larger, round coils, only that it is more practical for field use and that the slight loss in depth is more than made up for by the sharper response.

We have discussed the most practical BFO combination: a stable detector with approximately a 12-by-24-inch searchcoil. This coil should be operated four to six inches above the ground, regardless of whether the soil is mineralized, to help free the coil from ground effects and clear most obstructions. The question may be asked, "Why not try to operate closer to the ground and gain more depth?" When the BFO is operating at a smooth, even beat, the deep or faint signals are more easily heard at the suggested operating height. If you operate the coil TOO closely to the ground, you will have erratic or uneven audio response which will more than offset the advantage gained in lowering the searchcoil.

Let's discuss a typical search for money buried in a location such as an old homestead, ranch house, saloon, ghost town, etc., or in any area where there is probably an abundance of small metallic junk. Someone will likely say, "I can call my shots and tell the cache from the junk." Perhaps this is possible in some circumstances, but not always. Small junk could have been scattered on top of the buried cache by accident. Ten to one the cache will be in a can, kettle, jar with a metal lid, or in some other junk or ferrous container that would respond the same as all the other unwanted trash. This non-profitable trash will have to be removed patiently and slowly if you are determined to make a recovery. Answers to questions asked of any experienced, successful and truthful cache hunter will probably alert you to the fact that he has the same junk removal trouble. When one stops investigating all targets, the success ratio drops.

The BFO with large searchcoils is quite easy to use in trashy areas, BUT, it must be remembered, if you are passing over small junk targets you may ALSO be passing over the larger size container which lies at a deeper depth. It is electronically impossible to bypass small conductive targets lying close to large targets without some degree of detector response. The lingering doubt that haunts all treasure hunters is: "Did I use the deepest seeking detector available?" or "Did I pass directly over the treasure without knowing it?" The successful cache hunters continue to remove the small trash and perhaps that is why they ARE successful.

To sum up, the deep seeking VLF/TR types will produce the most depth in mineralized ground leaving the BFO type with large coils a second choice.

CACHE HUNTING WITH THE TR

To search for caches in non-mineralized zones you can use the TR. Try to obtain coils larger than the standard twelve-inch that comes as an accessory on some models. These extra-large TR coils are quite heavy, but they will penetrate mineral-free soil. Of course, you will NOT achieve the depth the VLF and PRG types can produce. Attempts to search in highly mineralized zones will meet with instant failure due to erratic detector operation, but you will at least be able to make your coin hunter TR double as a cache hunter under favorable conditions.

The extra-large searchcoils allow you to bypass many troublesome signals received from coins and small junk while engaged in the search for deep caches, large cannonballs, battlefield relics (*not* small bullets), old bottle dumps, and the like. Of course, ANY type detector that permits attachment of large searchcoils will perform admirably under such favorable circumstances. The VLF, PRG, and RF models will all produce approximately the same depth in mineral-free ground. The VLF will produce the *most* depth in mineralized ground, but, if you own a good TR capable of using extra-large searchcoils, you could expect reasonable depth penetration *in mineral-free ground*.

CACHE HUNTING WITH THE PRG

As mentioned previously, the PRG detector has almost perfect target identification and depth equaled only by the VLF and RF (two-box) type in non-mineralized zones. It takes two or more operators to use the large searchcoil that is available. It is impossible to be precise, but if the suspected cache site were in a salt water area, the PRG equipped with the large coil would probably detect more deeply than any other type detector and produce almost perfect identification of the target (the container in which the cache is buried). Of course, the PRG's use is limited to the search of areas relatively free of iron mineralization and it is quite costly, especially if the large searchcoil is included. However, cost is never a factor with professionals who consider only the possibility of success.

All salt water areas are not free of iron mineralization. Most ocean beaches are saturated with black sand (mineral) and are more or less negative. The tide comes in, wets the sand with salt water, and the area becomes conductive (positive) to all detector types except the PRG and PI. In low-lying marshy, tidal zone areas the sand never dries out. Here the PRG is rated tops in discrimination or target identification, equaled in depth only by the VLF type, provided the same size searchcoils are used.

Some people are confused when it comes to thinking about mineral salts that become conductive (metallic response) and mineral (Fe_3O_4, magnetic black sand) that produces a negative response. Both are considered "mineral," but the salts become conductive *only* when wet. The VLF type can be adjusted to operate mineral-free over any amount of negative black sand (mineral, Fe_3O_4). The ground control will simply zero out the magnetic black sand and permit full depth penetration. The PRG cannot be adjusted to operate on the black sand, but, as we said earlier, it has the advantage of target identification in salt water areas.

CACHE HUNTING WITH THE PI

This super-sensitive instrument will penetrate soils that contain a high degree of mineralization. It has the disadvantage of being overly sensitive to small ferrous iron (the same as the VLF type) and most models do not permit the application of large searchcoils. It is harder to use for pinpointing as the after-ring of the audio signal is very confusing in littered areas. The PI type detector is excellent in many circumstances, but it is troublesome and time-consuming to use around suspected cache sites in trashy or littered areas.

Deep cache recovery requires the use of large searchcoils. All models of the PI type are not versatile so take care to choose the model with large coil capability. In mineralized areas it is comparable in operation sensitivity to the VLF type. When the large searchcoil is used, it is very hard to narrow the target area and pinpoint accurately. Models permitting the application of large coils are *much more costly* than the VLF types and, considering that the PI has no more depth or versatility, it is hard to justify the added expense. It is imported from England and distributed throughout the United States. It is a quality instrument, but does not perform as favorably in applications such as cache hunting as do American-made detectors. If you desire a PI for cache hunting, we suggest you contact a U. S. distributor for more information. They give true and explicit information in regard to this product. We have used these PI instruments, and under certain circumstances they are effective. They produce false signals on extra hot or highly mineralized rocks that are "out of place" and do not have a separate discriminating circuit which will identify hot rocks. The oft-quoted remark, "Specialized instruments should be used only where they excel," holds true.

CACHE HUNTING WITH THE RF

One of the oldest type detectors is the RF (two-box). It definitely has a field wherein it excels: *pipe and cable location.* It is adversely affected by magnetic iron or mineral-salts (mineral content of search area). It produces more false signals than the VLF type as it responds to many of the earth's anomaly changes. It has the advantage of responding ONLY to targets that can be considered larger than a baseball. This characteristic permits the exclusion of response on most small trash items when cache hunting. If the soil is relatively free of mineral content, the RF will produce great depth and fast ground coverage, an added advantage when cache hunting anywhere.

The RF is difficult to use if the soil contains excessive mineral or if the location is in a rocky area. In such cases the RF must be detuned to the point where it loses its depth penetration and becomes almost useless. Do not attempt to use the two-box RF in salt-water zones or marshy areas as the depth is *drastically* reduced. The two-box also responds to both metal and mineral changes, leaving the operator at a complete loss as to correct target identification. It is highly productive in flat terrain where mineral is not abundant. Over the years, many large caches have been found with this type detector.

Many treasure hunters have used two-box detectors with success. However, since the introduction of the deepseeking VLF types, the two-box RF is losing its popularity. VLF types will detect smaller targets in mineralized rocks and soil, and they have the advantage of complete versatility in operation. The application of extra-large searchcoils on VLF types produces as much or more depth in mineral-free ground and *much* greater depth under highly mineralized conditions. Treasure hunters will hate to see the old two-box deepseeker retired as a cache hunter, but electronic facts cannot be changed.

CACHE HUNTING WITH THE VLF

Under excessively mineralized conditions, we recommend the VLF/TR type detector with large searchcoil. The VLF/TR will penetrate the mineral; the large searchcoil will free the super-sensitive instrument from response to some extremely small trash items, such as small nails, pieces of wire, tin, etc.

The VLF/TR will go deeper, operate perfectly, and outperform any other detector type, *provided* you employ the large searchcoils available for these deepseekers. Do not become overconfident and rely on the small size coils. If a cache is worth recovering, it is worth the small added cost of a large coil to ensure that you have the best and most efficient instrument at your disposal.

NOTE: The RF detector permits quick ground coverage, but it will not penetrate heavy mineralized background. Under these conditions the two-box RF models have been known to miss many cans, jars, and other small caches because the mineral content forces you to detune the detector. This also includes attempts to use standard TR (IB) detectors with large coils attached. In highly mineralized areas they must be detuned to the point where they lose *considerable* sensitivity.

VLF types should be used with the largest coil available; the *larger* the searchcoil, the fewer small nails and tiny pieces of

The authors are in a remote area in Mexico searching for hidden ore highgraded by silver mine workers. This mining district is in a highly mineralized (magnetic black sand) area and has been searched unsuccessfully by several professional hunters over the last few decades. However, the VLF/TR ground canceling detectors used by the authors ignored the magnetic mineralization and made ore and money cache detection possible.

ferrous iron you will have to dig. The larger coils on the VLF's are quite heavy, but they produce more depth.

Probably most cache hunters have left many deep caches that were beyond the detection range of the detector employed. Most professionals use the best detectors available and employ the large coil attachments. Even so, many caches, buried at a shallow depth but concealed under mineralized rocks, are missed. This can almost always be attributed to the detector's failure to penetrate the magnetic mineral rock successfully. The loot or relics are still there, awaiting the next hunter who is using a mineral-free-operation type of detector.

No doubt, many failures can be attributed to the person who may be thoroughly experienced as a coin hunter but inexperienced as a cache hunter. Because he has full confidence in

The authors look down upon the famous crossing where in 1877 Nez Perce Chief Joseph led his people across the flooding Snake River into Idaho. A large Indian gold cache was buried near here during the Nez Perce Indian wars in the late 1870's.

The authors take an evening to rest along the famous Snake River in Idaho, U.S.A., a few miles below Hell's Canyon Dam. They discuss the activities of the next few days with Sandy, a jet boat pilot who took them several miles up the river to a location where the men searched for a large Indian gold cache which was buried during the Nez Perce Indian wars in the late 1870s.

his detector for coin hunting, he overestimates its penetration on deeper objects. An erroneous interpretation may be arrived at by his having recovered a tin can or larger object at great depths while coin hunting. No consideration was given the terrain which was easily accessible with his coin hunting detector and small searchcoil, and no allowance was made as to the presence of a small amount of mineral content.

The cache hunter does his research and comes across information leading to a ghost town or suspected treasure site, perhaps an old Spanish mission, miner's stash, site of a stage coach robbery, Indian cache, etc. He may spend considerable time doing all research possible, enlist the aid of a partner, take all kinds of camping equipment, and spend a few hundred dollars for a weekend expedition. After this expenditure of time and money, he generally neglects the most important tool necessary for recovery of the cache — the correct *type* of detector, one capable of utilizing *large* searchcoils. If the area is mineralized (and almost always it is) a detector type that will *penetrate* mineralized rocks or ground is required. Perhaps his prized detector is a standard TR, so popular for coin hunting; perhaps it is a discriminating type capable of utilizing only smaller searchcoils. No matter... he is almost helpless. In many magazines you will see pictures of groups searching in isolated areas that probably required many days of preparation and much expense to reach. You will notice oftentimes that the searchers are using small eight-inch or medium size searchcoils. These treasure hunts generally produce a few old coins or ancient relics but seldom a cache, especially one buried in rocky mineralized areas.

If you want experience and want to convince yourself of the problems involved in cache hunting, obtain a gallon bucket or can and a large slab of highly mineralized rock approximately four to eight inches thick and two or three feet wide. Dig a very SHALLOW hole and just barely cover the gallon can. (It does not matter whether the can has money in it... a metal detector only "sees" the outside of the can.) Slide the large rock slab over the buried can; tune your detector to the ground mineral content (not over the rock); and pretend you do not know the cache is under the rock. Start your search and pass over the large cache. You probably did not get ANY indication; if you did it was very slight. Consider that a gallon can will hold approximately $27,000 in gold at the old price and approximately $1,200 in silver dollars. (Understand this large a cache is not common as many consist of only a few thousand dollars at most.) We are talking about a SMALL can (for example, a tobacco can) or quart container, perhaps an old fruit jar with zinc lid or maybe saddlebags with a few handfuls of gold coins in them. Consider what will happen when the suspected cache consists of only a few thousand dollars in gold and may be buried anywhere from one foot to three feet deep. Large iron chests and kettles are not really TOO much larger than the gallon can and, if they are concealed under mineralized rocks or buried at arm's length in mineralized ground, you may now understand WHY you HAVE NOT been finding them.

Professional cache hunters realize all this and make allowances for the condition of the search area and the fact the cache may be deeper or smaller than anticipated. They take all the precautions they can, such as attention to *deepseeking* instruments using the *large* searchcoils. The simulated cache under the rock slab and in mineralized ground would also have been very difficult (if not impossible) to detect with a BFO type equipped with large searchcoils. The VLF types provide your

Roy Lagal, somewhere in the state of Texas, U.S.A., tries to remove a few reluctant rattlesnakes from a small cave entrance. This cave had been dynamited closed immediately after a large assortment of outlaw loot was hidden there.

best chance for recovery in rocky or rough terrain where mineralization is present. When magnetic iron oxides are present (approximately 90 percent of the world soil contains iron minerals), the VLF type sometimes responds to small out-of-place hot rocks. As mentioned previously, this will only occur when the rocks are in rather close relation to the searchcoil and is known as an antenna or near field effect. These excessively magnetic mineralized rocks merely unbalance the tuning and produce a positive or metallic response. The PI type has the same problem due to the fact that it does not, or perhaps cannot, utilize a standard circuit capable of metal versus mineral identification. You would have to dig EVERY INDICATION! Any VLF type without an incorporated TR circuit capable of identifying metal versus mineral in these highly magnetic zones is useless for cache hunting. You would spend all your time digging and investigating hot rocks.

All VLF types that incorporate a low frequency TR circuit can instantly identify and reject the hot rocks. When you use the VLF/TR type for cache hunting, adjust the TR discrimination control to zero. (Any amount of discrimination might cause you to pass over an iron kettle or box and would be useless in this application.) Use the largest coil you have available and, switch to the VLF mode of operation to begin searching. Depending on the searchcoil size, operate the searchcoil from six to ten inches above the ground surface to prevent the supersensitive circuit from detecting a few of the extremely small targets that could *not* possibly be a cache.

You receive a target indication! Determine the exact center of the target by detuning the detector. Use the push button or master control switch method. Move the searchcoil to one side of the target maintaining the same height of operation as used during the search. Switch to the TR mode. (Since you have set the discrimination previously to zero, the circuit will ONLY indicate whether the target is of some metallic content or simply a hot rock with a high magnetic content.) Do not forget, always retune the detector each time you change modes. Move the searchcoil back over the exact center of the target, making sure the audio response is loud enough to hear. If the audio decreases slightly, or stops completely (as it probably will), the target is only a hot rock or mineralized spot. If the sound continues to respond at the same level or increases, you have determined the target is a metallic object.

Further use of the discrimination control to identify the target is generally useless in cache hunting as the detector is responding only to the container, usually made of iron. In certain situations where you are looking for a large non-ferrous target (glass jar of coins, silver in leather saddle bags, etc.), you could continue to increase the discrimination amount until you were convinced the target was not ferrous iron. However, this is seldom the case and investigation of ALL targets usually pays dividends. Your primary objective is ONLY to determine if the target is metal, NOT what type of metal it is. The container could possibly be some type other than indicated by the treasure lead story. Thus, you see, the VLF type containing a TR circuit can instantly identify the mineralized hot spots and make cache hunting a pleasure.

When the VLF types first became available on the commercial market, it was thought it was impossible to cache hunt in heavily littered trash areas. Of course, a less sensitive type of detector would bypass the small trash, but *also miss the cache*. With the introduction of the instant retuning methods available on most quality VLF detector types, cache hunting in trashy areas is both feasible and productive. When a multitude of targets are located in close relation to each other, simply keep detuning the circuit by use of the instant retuning method until you have isolated each target. They might be only small trash objects, but it pays to investigate every possibility. Now you can understand why successful cache hunters accept the fact that they must dig worthless trash much of the time. Cache hunting takes patience and perseverance. Cache hunting has its problems, just like coin hunting. One problem is that if the detector has sufficient sensitivity to penetrate the earth deeply for a small target, it will also have greater sensitivity to a small target that is closer to the searchcoil.

NEVER, and again we say, NEVER pass a suspected treasure site just because someone tells you it has been worked before. More treasures are missed than recovered, all due to the depth of burial and mineralization content of the ground. Consider old parks where coins are found; the number of years they are hunted by thousands of coin hunters; and the fact they

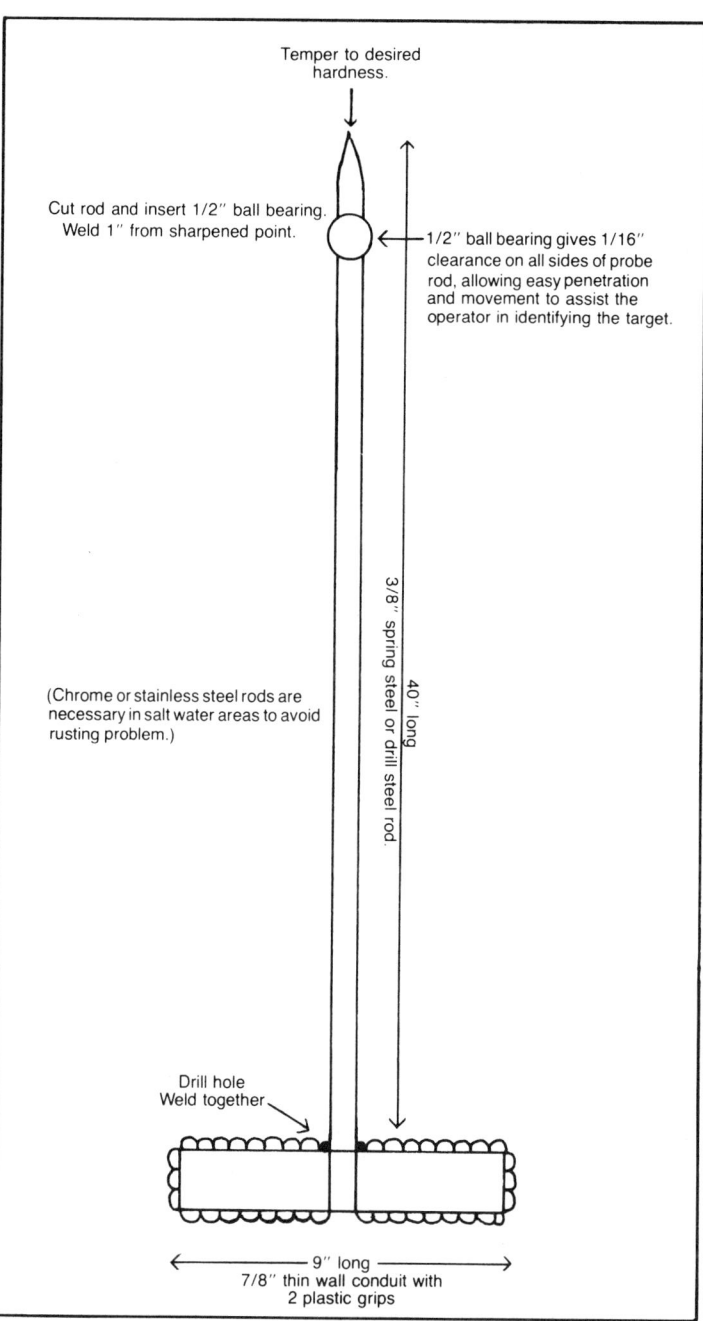

never become completely hunted out. The same is true of suspected caches. When better detectors come along, these old sites should always be promptly reinvestigated. If you do not, someone else will. Many recent recoveries have already been made from old locations since the introduction of the mineral-free operation, deepseeking VLF/TR type detectors.

USE A PROBE FOR CACHE HUNTING

Always carry a time-saving steel probe to use where the soil permits. If you think the response from your detector indicates a target large enough and deep enough to fit your idea of the suspected cache, probe the spot before digging. Experienced operators can probe carefully and determine what the target is. A glass jar would be easy to identify. If the probe contacts a flat piece of tin and you forced the probe on through and felt nothing else, the target would probably be just that... a flat piece of junk tin. If you contacted a tin can and the probe penetrated so that you could tell there was something in the can, you would want to dig it. Many cache hunters who use probes become so proficient with them that they can feel a newspaper when the probe passes through it.

A special kind of probe rod is required. A picture and the instructions for making one are included herein. The probe has a steel ball-bearing welded onto the point end to permit the probe to move up and down with no restrictions so you can define more clearly what the underground object is. If the ground is rocky or hard, a probe is not helpful. In that case you will just have to dig all indicated targets that you judge deep and large enough to be the suspected treasure cache.

You may wish to use a backpack when in the field, as many of us do, either en route to a location or coming from one. It helps avoid drawing attention from the curious because we appear simply as hikers just enjoying the scenery... plus the pack is a convenient way to carry equipment and supplies. Large searchcoils will generally fit into the large backpacks, as well as small shovels, detector, probes, and the tools necessary to make an average recovery.

THE CACHE HUNTER... A PROFILE!

There are all kinds of treasure hunters, and then there is the cache hunter. He or she is generally not interested in a few relic items and a couple of coins which probably would not even pay for the gas. The cache hunter is interested in finding loot — a lot of it! Of course, the hunter does not find such caches everyday. The beginner must realize this and not become discouraged. However, there are millions of dollars stashed in the ground and, if you persist, sooner or later you will hit a cache. A lot of people have become wealthy from pursuit of this interesting occupation.

The genuine true cache hunter is definitely a separate breed in the treasure hunting fraternity. This person is seldom seen among weekend hobbyists, those who hunt for coins and relics just for the fun of it! There are many highly successful cache hunters who never never coin hunt, preferring to spend all their time in pursuit of larger, more profitable finds. None consider coin hunting as a hobby beneath their dignity. It is just that small object searching never appealed to them. (This is true of the big game fisherman who spends his entire life in the

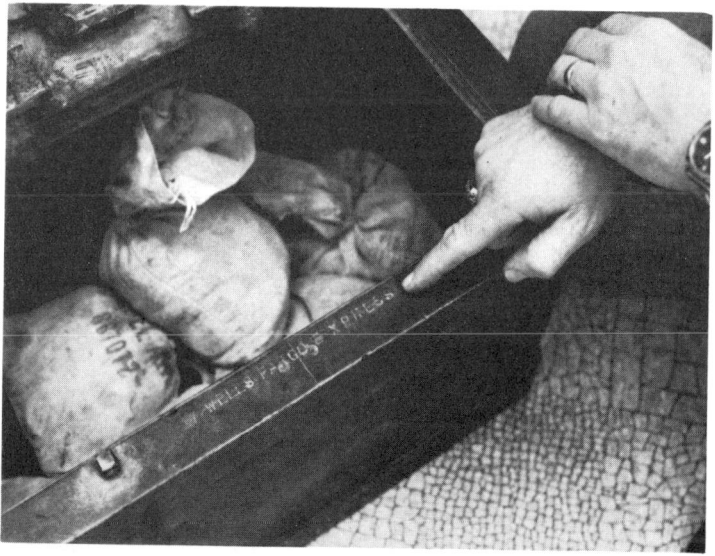

A true story involving recovered outlaw bank loot and this Wells-Fargo strong box was printed in a national treasure hunting publication. The cache was found in Oklahoma by a well-known treasure hunter.

search for record-breaking marlin or sailfish.) Most, if not all, real cache hunters spend much of their time in research, seldom mentioning their occupation to anyone other than perhaps another professional. Their actual expenses are considerable as research sometimes demands extensive travel. They may occasionally have to pay a rather large sum of money to obtain information. Perhaps they may even offer to share the cache on a percentage basis, a very common practice, especially for permission to search on private property. Research is very time-consuming, and it is advisable to work on more than one project at a time. Due to various soil conditions it is often necessary to purchase special types of detectors. Patience, as well as financing, is required.

If you do not want to bothered by the curious or those not actually entitled to a share of a recovery, maintain a low profile. DO NOT put your trust into a *verbal* agreement with a landowner and DO NOT leave an open hole after you have removed something.

It is only fair to have an agreement in writing in proper legal form. Generally, you can trust a cache hunter who makes his living in the business. For one thing, he cannot afford to have his name besmirched in his trade and, for another, he has handled found money and does not get excited and perhaps think of trying to take your share.

You should seek permission before entering private property. If a cache was buried long ago there may be a problem in determining the legal ownership of the property where it is buried. Who does it belong to? The land belongs to an individual, the state or the federal government. According to most state laws governing treasure trove, it belongs to the finder; however, you may be guilty of trespassing if you do not secure permission to search. The cache does not become treasure trove until it is found. If the find must be turned over to the authorities to decide ownership, the case may be tied up in court for a long period. When all is said and done, lawyer and court costs may take up the biggest share.

Some states have enforced trespassing laws and some have not. Consider your situation wisely! You may wind up in court if you do not first obtain permission to search. It would be difficult to go forth in the wide open spaces and fail to break, unknowingly, some particular law. Respect all property rights. If you treat property owners with courtesy and friendliness, they will more than likely be cooperative.

PAY TAXES ON RECOVERED CACHE?

Do you think you can find $50,000, buy a new car, take a long-waited vacation, etc., without the extra income being noticed? Hardly! So declare the extra cache money, pay your fair share and rest easy. Of course, you may owe nothing until you convert the value of the cache into spendable money; then it becomes taxable income. If you are among the many professionals who make their entire living at this, probably you will be able to deduct expenses incurred while engaging in search and recovery. Many states have explicit laws governing treasure trove and can claim part or all of everything. Check with your attorney!

For those of you who would like to delve further into the fields of research, recovery, tax problems, equipment advice, and gain a thorough insight into the pro's and con's of profes-

Charles Garrett and A.M. VanFossen demonstrate deep seeking, cache hunting metal detectors to Mrs. Pancho Villa. Pancho Villa is believed to have buried vast treasures in several places in both Mexico and the United States. These treasures have been searched for by countless cache hunters. It is believed that many of Pancho's treasures have never been found.

sional treasure hunting, purchase a copy of *Treasure Hunter's Manuals #6* and *#7* by Karl von Mueller and *Professional Treasure Hunter* by George Mroczkowski, published by Ram Publishing Company. These are three of the most complete and most informative books on these subjects.

CACHE HUNTING SUMMARY

Most of us have spent years searching for suspected caches on rocky hillsides, canyons, rock bluffs, talus slides, and river banks with gravel concentrations — areas far removed from recent habitation that probably contain only small amounts of trash, though they may be highly mineralized. Mostly we have relied on the BFO and RF (two-box transmitter-receiver) type detectors; however, a few successful hunters have employed the early-day horizontal loop TR detectors in certain areas. The fact that huge rocks and canyons are mentioned does not mean all rock contains the magnetic iron that is so disastrous to effective TR operation. A few isolated areas in the bluff or canyon section of Utah, Colorado, Arizona, New Mexico, and elsewhere are of limestone or sandstone composition and may contain no magnetic oxides.

As mentioned previously, probably ninety percent of the entire earth's surface contains iron oxides to some degree. The earth is thought to be solid, consisting of many minerals and metals, but only heavy magnetic iron oxides create problems for the metal/mineral detector user.

Often, cache hunting failures can be blamed on ground minerals and on the reluctance or inability of a detector dealer to explain that a model he sells does not have deep detection capability or the versatility required for cache hunting. In certain states some particular type of detector may have achieved widespread popularity. This may be because local dealers stock this specific model or brand almost exclusively. A few dealers are only sales motivated or perhaps just plain inexperienced and do not bother to explain that another type of instrument would be better suited for your particular job requirement. This combination of conditions sometimes leaves the cache hunter who wishes to become successful no place to turn, except perhaps to knowledgeable books on the subject. Even some highly advertised books are not the final answer.

They may be written to promote a particular brand of detector, or they may be written by an armchair treasure hunter who has had limited field experience. All detectors should be explained and presented in an unbiased manner. Field tests are always the buyer's best protection and guide.

CHAPTER 12

Searching for Coins

It is estimated that 90 percent of all detector owners occasionally search for coins. This is why coin hunting is regarded as the most popular phase of treasure hunting.

Since the search for coins is simple and easy, all family members from small children to adults can participate. To be succesful you need a good quality metal detector and a bit of research to locate good coin hunting spots.

In this chapter, we discuss detector selection and usage. We will explain all detector types and how they are suited (or

Bob Podhrasky, internationally known engineer and metal detector circuit designer, demonstrates the scrubbing technique. Often when scanning in the TR discrimination mode, this scrubbing technique allows the operator to achieve the best results. The searchcoil is pressed firmly on the ground and scanned or slid across the ground. This keeps the searchcoil at a fixed distance in relation to the ground, thus eliminating much of the erratic response which sometimes occurs when operating over mineralized ground.

not suited) to coin hunting. We will analyze major detector features, such as optimum searchcoils, Faraday shielding, sensitivity, discrimination, and ground canceling circuits.

SEARCHCOIL SELECTION

There are many factors to consider when selecting a metal detector. The eight-inch searchcoil is the most practical TR searchcoil size. With the BFO, the three-inch, five-inch, and six-inch coils will be the basic choices. Many discriminator detectors come equipped with only one small searchcoil that is attached directly to the control housing circuit board. These coils cannot be easily changed or removed. The PI responds well with the eight-inch coil. The PRG is usually equipped with an eight-inch coil. The RF does not respond to single coins. The VLF types are generally designed with six to eight-inch searchcoils. The seven to eight-inch is the most practical and best producer on small targets, provided it is designed to pinpoint accurately. Faraday-shielded searchcoils are very important on any type of detector because interference caused by weeds and grass will sometimes distort or blank out faint signals from deeply buried coins.

COIN HUNTING WITH THE TR

The optimum TR setting is as follows. Set the tuning in the metallic mode of operation. Place the searchcoil flat on the surface of the ground or grass. Adjust the tuning until you hear a faint audio sound and move the searchcoil in a side-to-side motion. Then, as you walk forward, continue the side-to-side scan, overlapping each sweep slightly. Some operators tune the TR in the null or quiet audio zone. While this does make for quieter and perhaps less irritating operation, it also increases the chance of missing deeper coins. If the TR is tuned silent there is no way to tell when minerals or drift force the tuning farther back into the null zone. This causes loss of a portion of the detector's sensitivity.

It is not mandatory to operate the TR searchcoil flat on the ground's surface. You may operate one or two inches above the ground. You will, however, gain a couple of inches of depth with the ground scrubbing action of the coil and you may get better

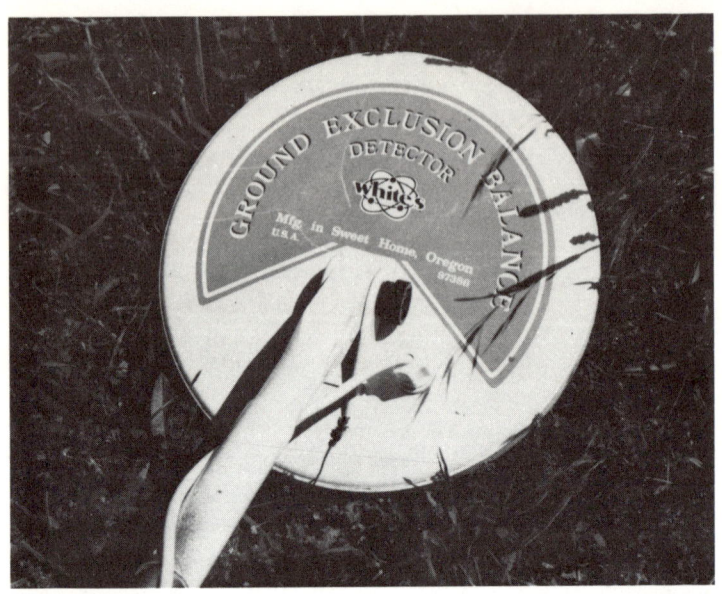

Manufacturers devote a great percentage of their time to developing and producing coils that will withstand the tremendous punishment that is asked of them, while still maintaining perfect depth penetration and operation. You'll note these two searchcoils are constructed with short plastic rods which connect the searchcoil to the upper metal rod portion. This type construction is necessary in today's high performance equipment. It prevents the metal rod from interfering with the detector operations. Rugged submersible searchcoils are as vital design characteristics as any other detector features.

Jim Alexander of Pasadena, Texas, well-known and widely traveled treasure hunter, searches for coins and jewelry on a Texas coast. The conductive salt water beaches often make it difficult to achieve 100 percent accuracy. Most beaches contain not only conductive salt but also magnetite. With practice, however, operating problems caused by these substances can be overcome, rewarding the persistent beachcomber with many valuable finds. Stop by to see Jim at his shop at 21 Spencer Highway, S. Houston, Texas 77587.

depth on single coins if you keep the detector properly tuned and slightly up in the tuning range.

COIN HUNTING WITH THE BFO

BFO operation differs slightly from that of a TR. BFO searchcoils operate best at approximately one to two inches above the surface of the ground, depending on ground mineral content. Set the tuning in the metallic mode at a frequency of from twenty to sixty beats per second. The speed setting may vary according to individual preference. Some like the moderate motorboating sound; some like faster tuning. The faster speed is *absolutely* necessary with BFO high frequency detectors or with poorly designed circuits that are highly erratic and unstable. You will find that the BFO's constant sound feature makes a signal harder to distinguish than the quick response of the TR, but with practice you will be able to hear even the smallest increase in beats.

BFO depth may be somewhat less on small coins than that of a TR, but it is easier to use over mineralized and uneven ground. Fully shielded searchcoils are very important on ANY type detector, especially in the search for single coins. Deeper coins produce fainter signals, and grass and weed interference produced by unshielded searchcoils garble and destroy faint signals, making them barely, if at all, distinguishable. BFO searchcoil choices will include the independently operated dual coils which are very popular because they allow you to conduct your search with the larger coil, then switch to the smaller, inside coil to pinpoint a signal. This arrangement increases the amount of search area which can be covered. With proper practice you can become quite an expert with independently operated dual coils.

The speed at which you sweep your searchcoil is dependent upon the detector's tuning frequency and your preference. If your detector is tuned at a slow motorboating sound, move the searchcoil slowly and keep it level with the ground. If your instrument is tuned at a moderate or fast motorboat sound, then you can move the searchcoil faster. Moving the coil TOO slowly will fail to achieve the audio "break" that is necessary for deep coin detection. Too fast a pace will cause you to miss coins. As with any detector, practice is required to become proficient in locating coins.

COIN HUNTING WITH THE PRG

PRG detectors produce almost perfect target identification and sensitivity compares with the VLF and PI types. Most detector types are affected by wet mineral salts, but the PRG operates perfectly around salt water areas. When salt becomes wet, the solution becomes conductive and responds as metal. Since the PRG is not affected by that condition, it can be very successfully used for coin hunting. It will NOT, however, operate adequately in iron mineralized areas. It is adversely affected by even small amounts of magnetic black sand (Fe_3O_4). The PRG is rather heavy, about seven pounds, and requires some operator skill to use. It might be considered overly expensive, but not for selected coin hunting in specific locations, such as in salt water areas where valuable Spanish coins may be found. If you are coin hunting in areas free of magnetic content, the PRG will produce almost perfect identification on targets and achieve fantastic depths. For best results follow the manufacturer's instructions on tuning and operation.

COIN HUNTING WITH THE PI

The PI is much lighter than many detector types and features a searchcoil that is completely unaffected by grass, weeds or salt water. The response of the PI is a bell-like ringing which makes pinpointing difficult. You must occasionally detune or reduce sensitivity in order to pinpoint accurately. PI's generally feature earphones, but a small outside speaker is simple to add. There is no zero control to adjust for mineral content, but it does not need one. Merely turn the tuning control on and advance it forward until the point is reached where the bell starts to ring. Back it up slightly, and you are ready for operation.

Start your search by holding the searchcoil approximately two to four inches above the ground, taking care not to vary the operating height too greatly. This is important as the PI will respond positive (metallic) if brought into contact with or placed in close proximity to mineralized ground. Keep the searchcoil height approximately level. One difficulty is pinpointing the indicated target in relation to or among other small metallic objects. The instrument's super-sensitivity to small ferrous (iron) targets sometimes makes operation difficult. Expert operators achieve a certain amount of discrimination with this detector, but its operation is best limited to certain non-trashy coin hunting areas.

The PI is an excellent coin producer on salt water beaches if ferrous trash is not abundant. In this situation, the PI will operate easier than a VLF/TR type; however, the PI is not capable of discrimination and the VLF/TR is. Your choice will be influenced by the amount of conductivity present in the salt water area and the amount of small, ferrous iron metal pieces present in the sand. Perhaps the PI will someday be produced in a lightweight, full-time coin hunting discriminator configuration. Many complicated and almost unworkable versions of this type detector are constantly being experimented with and offered in various styles and models. Stick with well-known manufacturers who are capable of backing up claims for their products.

COIN HUNTING WITH THE VLF

VLF detectors are excellent for locating coins or small metallic objects at great depths. However, without some type of discriminating ability these deepseekers are almost impossible to use in trash or littered areas. They are extremely super-sensitive and will detect tiny ferrous and non-ferrous objects so small the eye can hardly see them. They will detect nails at fantastic depths and consequently are in constant demand by loggers, sawmills, and surveying crews. They will operate perfectly in highly mineralized areas with very little loss in depth and find deep coins that many other detector types cannot. If the area you intend to search has been hunted previously, these deepseeking VLF types will produce the best results.

Earth iron mineralization can be completely "balanced" or adjusted out, permitting the searchcoil to be operated at almost any height. This enables the operator to conduct his search over mineralized river gravel, in areas containing concentrations of black sand, and in other difficult places where the standard TR is almost impossible to operate and where the BFO type experiences a definite loss in depth. Simply adjust the VLF searchcoil to the ground by use of the ground or terrain control according to the manufacturer's instructions. (Do not confuse this control with conventional "ground" or "mineralized" adjustment features on standard TR's.) You may then adjust the tuning to the desired level of sound and proceed to search.

While attempting to pinpoint, you may also achieve a certain amount of discrimination. Pass the searchcoil forward and backward and side-to-side over the target. If the target is an elongated ferrous object, such as a nail or a small piece of iron longer than it is wide, you will probably receive a double or "blip blip" response. If you hear only a single "blip," the target is either non-ferrous or perhaps a round iron object. This method of partial discrimination lacks total accuracy, but, when interpreted with operator experience, it can save a lot of useless digging.

Frank Mellish of London, England, a charter member of the International Treasure Hunting Society, takes time from an Australian outback gold hunt to search for coins. Shown in the lower photograph are some of the nuggets that have recently been found by modern-day electronic prospectors using VLF/TR metal/mineral detectors. Nuggets such as these were missed by the early-day miners in desert gold fields because lack of water greatly slowed down their operations. Modern detectors will find even extremely small nuggets in such areas as this, in spite of the highly mineralized ground matrix. Frank has accompanied Charles Garrett on several treasure hunting and equipment testing trips throughout Europe, Australia, and the United States. Frank is one of the best-known and most successful European treasure hunters and is one of the key men responsible for the sudden, rapid growth of treasure hunting in the United Kingdom.

The authors search for coins at a roadside park. At this location, they found the automatic mode of operation while searching the in TR discrimination mode gave the best results.

Searchcoils should generally be operated at a height of at least two inches above the ground. When you receive a loud signal, you may find it necessary to back off or to detune the sensitivity in order to pinpoint the target. You can now understand the problems that you may encounter in coin hunting in trashy areas with super-sensitive instruments that do not have TR discrimination capabilities. Small searchcoils, generally seven to eight-inch diameter, are the most practical for single coins because they are lightweight and allow for quick pinpointing.

Different manufacturers use different configurations in their searchcoil designs. The co-planar design is the most popular and it is easy to pinpoint with it. The target is easily centered and the flat bottom of the co-planar coil allows the operator to move the coil from side-to-side over the object without unnecessary variation in operating height. This pinpointing method can be accomplished while the searchcoil is in direct contact with the earth. In some cases, this direct contact with the ground is necessary during the identification or discrimination procedure. Very little electrical interference is experienced with co-planar type coils, but close proximity to television sets may cause some fluctuations in a very low frequency circuit.

The co-axial searchcoil is commonly known as a "stacked coil." The transmitter and receiver coils are placed or "stacked" one on top of the other. This procedure virtually eliminates all electrical interference caused by television, power lines, fluorescent lighting, etc. Many operators prefer co-planar type searchcoils because they do not like the appearance of the co-axial coil. Also, they find the co-planar is easier to maneuver. On the other hand, co-axial searchcoils have near-perfect operating characteristics and are definitely preferred by some operators.

A few manufacturers wind their co-planar VLF/TR searchcoils in such a manner as to achieve a certain amount of discrimination while operating in the VLF (mineral-free) mode. This type of audio discrimination is interpreted entirely by the operator and is perhaps fifty to ninety percent efficient. This method of discrimination is limited solely to ferrous target rejection. Foil and aluminum pulltabs must continue to be rejected by the TR discriminating circuit. This operator-interpreted discrimination does, however, allow for fast shallow depth operation in trashy areas because most shallow iron targets can be quickly identified. The method is interpreted as described below.

Targets with magnetic content (ferrous) respond much *wider* than searchcoil width. The audio response on a coin or non-ferrous target will start when the searchcoil reaches the exact edge of the target and increase quickly to a peak. The response then drops off quickly as the coil is moved away. Identification of a large, non-ferrous target can be accomplished by this method if you will keep the searchcoil far enough away so that the target's signal does not fully saturate the audio circuit. You may easily practice this method by using a coin and a bottlecap of equal size. The results will amaze you! If your detector is capable of this kind of discrimination, lay it flat on a wooden table, switch to the VLF mode, adjust the tuning to a slight audio, and move the bottlecap slowly across the searchcoil surface, approximately three inches from it. Notice the response will start slowly climbing before the bottlecap comes beneath the coil's surface. Continue across with the cap and

Allan Cannon, one of the treasure hunting scene's youngest, most popular and well-known treasure hunters, spends almost all of his time coin hunting and electronic prospecting, but occasionally he finds time to search for battlefield relics. He has spent a great deal of time searching for treasure in Idaho, Montana, Washington, Oregon, Texas, and Canada. He is shown here with a rotted leather coin pouch that he found. The pouch contained several old coins, one of which he is displaying. Allan is frequently a first place competition hunt winner and has filled a large display area with trophies that he has won. Photo credit Darrell F. Clark, Clarkston, Washington.

notice the sound fades slowly. This is the wide response we refer to. Test with the coin at the same exact distance from the searchcoil. Notice response does not start until the coin is approximately even with, or under, the coil's edge. Move the coin on across and away. The sound drops quickly.

Practice this same technique on larger targets. Hold the target much farther away so that the signal is about one-fourth to one-half strength. The large iron target will begin to respond long before the edge of the searchcoil is reached. Try the same test on a large non-ferous target (aluminum, brass, copper, etc.). Notice the response will begin quickly when the coil's edge starts over the target. It will drop off suddenly when leaving the other side. A small amount of practice will quickly convince you that you can identify at least fifty percent of all targets. Remember, if the target is too close and causes the audio to responde WIDE OPEN, you must raise the searchcoil until the signal is about one-quarter to one-half strength. NOW make a slower pass over the target. Anything that responds WIDER than the searchcoil is of magnetic content (iron); anything that responds NO WIDER than the coil width deserves further investigation. The VLF/TR type detector may then be switched into the TR discrimination mode and the target correctly identified.

There are many different methods by which you may utilize your low frequency TR discriminator circuit. We will describe a few, and your own experimentation and experience will undoubtedly disclose many more.

VLF/TR searching where rare and valuable coins have been discovered can be very rewarding. In such areas where one does not mind going more slowly, results can be very profitable. Switch the operation mode to VLF, adjust the tuning control to slight audio, balance or adjust out the effect of ground mineralization, and start searching. (This method will insure that you MISS NOTHING within the range of your particular VLF detector.)

As discussed previously, you may be able to bypass many worthless iron targets by discerning the width of the signal. Do so, and when you believe you have a good target, drop the searchcoil to the ground *directly over the target*. Press or flip the retuning button to retune the audio to that target. Move back and forth over the target until you have pinpointed it exactly. Slide the searchcoil sideways until the edge clears the targeted spot. Switch the VLF mode switch to TR discriminate position (you should already have a preselected amount of rejection set in the TR discriminate control). If necessary, press or flip the retuning button or switch to retune the detector. Slide the searchcoil slowly back over the spot where the target lies. If the sound increases, dig up the target. If it decreases, you may wish to leave the target in the ground. You may have to slide the coil back and forth and use the retuning switch several times to get an accurate audio response. A slight downward pressure applied on the searchcoil to prevent erratic audio response from ground minerals may be necessary, but you can be sure if your VLF type detected a deep coin missed by previous hunters that the TR discriminate mode will identify it.

Many operators will adjust their discrimination control differently when in a supposedly worked out old coin area. Some operators will adjust the rejection to exclude only nails, wishing to miss none of the tiny, marginal coins with slight pewter or iron content. Others will adjust for bottlecap rejection so as to miss none of the small rings and nickels. The less rejection you use, the *greater chance you will have to get deep coins missed in previous searches.* The experience you get in different areas will guide you. If the VLF/TR does not contain a multitude of unnecessary controls and if the control switches are easily accessible, you will be able to accomplish the above described operating procedures in a few seconds.

The results obtained from using the VLF/TR type detector for coin hunting in trashy areas depend on many things, such as: simplicity of controls; whether there is manual push button tuning, automatic tuning, or automatic mode switching; how accessible the controls are; and how quickly they can be operated.

Select a very trashy area, such as a picnic grounds or roadside park. Adjust the discrimination control to reject pulltabs; turn on the detector; adjust the tuning to achieve slightly more audio than usual; switch to the TR discriminate mode and operate the detector with the searchcoil approximately two to four inches above the ground. If the mineral content makes operation erratic, decrease sensitivity control until you can move the searchcoil in slow sweeps and maintain a fairly steady audible response. Listen carefully, for the bad targets will cause the audio to decrease and the good ones will cause only a slight increase due to the reduction in sensitivity. After some practice, you will find this method rather easy and highly productive on shallow coins. You will miss the small rings and nickels because of the discrimination setting. There are many trashy areas which are almost untouched, and this method can be highly productive if you have patience and practice with your detector. Some operators adjust the discrimination control to eliminate only nails. Many times, this prevents the circuit from decreasing and much of the trash is shallow so quick retrieval and identification of bottlecaps and pulltabs is possible. This procedure gives better response and depth on coins. As usual, experimentation and experience will guide you. If you own one of the VLF/TR type detectors that has the master control switch mounted in the handle, you will probably find it just as fast to search in the VLF mode and then to pass back quickly over the target in TR discrimination mode for identification.

Many VLF models vary greatly in the speed with which they may be operated. Try to choose a fast operating VLF with full TR discriminating capabilities.

Discrimination while operating in the fully automatic mode is certainly possible with the very low frequency VLF/TR's. In automatic tuning there is a rebound or overshoot problem of which you should be aware. All audio and meter responses, either positive or negative, will be followed immediately by a signal of opposite polarity. In other words, if the detector is set to TR discrimination and the searchcoil is passed over a junk target, the audio and meter will decrease suddenly; the audio and meter will increase sharply as the searchcoil glides on past the target. This will sound to you just as if it were a good, positive signal. It is, however, only the overshoot signal produced when the fully automatic tuning is used. In the same manner, when you pass over a good target, the audible (and meter) response will increase and then fall back sharply.

You should practice until you become familiar with this characteristic. Select a bottlecap and a coin. Place them on the

Probably 90 percent of all detector owners hunt coins at one time or another. A large percentage of all treasure hunters search for coins almost exclusively. Coin hunting is a very popular and rewarding activity which allows family members of almost any age to participate. It is estimated that approximately forty percent of all coin hunters are women. Several manufacturers produce lighter weight instruments and armrests for heavier coils. Armrests make the use of larger coils less fatiguing.

ground about a foot apart. Pre-set your discrimination control to reject the bottlecap. Turn the detector on, select the TR discrimination mode and adjust the tuning control to produce a good, audible tone. Now switch to the fully automatic mode of operation. Hold the searchcoil approximately four to six inches above the ground and start your search pattern. Move the searchcoil over the targets at a reasonable speed, taking note of the different responses. Differences will be easily distinguished. (It may be necessary to reduce the sensitivity according to the amount of magnetic content of the ground. To do so will help prevent excessive erratic operation.) Continue your practice sessions and gradually move into an area where extreme trash conditions prevail. Of course, each individual is different and some will have more difficulty than others in recognizing the different signals, especially at faster operating speeds. However, this method is used successfully by many experienced operators in high trash areas and produces good results on shallow targets. It is simply a matter of learning to recognize and sort out the false overshoot signals from all the good signals. Moving at a constant speed will help you to achieve better accuracy.

There is no end to the different VLF/TR discriminating methods which are practical and can be successful. VLF/TR detectors have only recently been developed for use in coin hunting, and operators who take the time to learn the detectors fully are finding literally dozens of new applications and search methods.

Search where the coins can be found! Consider the fisherman who finds a likely looking spot and very patiently fishes there, but with no success. He will then move on to try another spot. Many detector operators, believe it or not, will start in an area and continue to find and dig small trash WITHOUT FINDING A SINGLE COIN, complaining all the while that all they can find is junk. If you are detecting small pieces of junk you would certainly detect a coin which is larger. Thus, you are NOT MISSING any coins: there are simply none there. The experienced coin hunter quickly moves to another spot, tests it, and continues this procedure until success is found. Keep this point well in mind. Regardless of the super-sensitivity and versatility of your detector, YOU SIMPLY CANNOT FIND COINS

WHERE THEY DO NOT EXIST! Never place the blame on your detector; accept the fact that the area is barren and move on.

When using any type of detector that is affected by the magnetic soil content of the earth, remember one electronic fact that simply cannot be ignored — ANY conductive trash that is between your searchcoil and, say, a coin will affect the searchcoil's electromagnetic field. This may result in your deciding that the target was junk. Not so, obviously, in this case! Practice by distributing small conductive junk targets at shallow depths with a coin placed more deeply. You will quickly notice that the interference from the smaller, but closer, junk targets reduced or distorted the coin signal. Sometimes, it may pay large dividends to go ahead and dig ALL trash targets so that you can reach the deeper and more valuable coins.

To summarize, we recommend that all coin hunters consider their locale and surroundings. Those who search parks

Coin hunters, well-trained in the use of their detectors, are able to go back to coin hunting sites which have been worked and reworked and still discover handfuls of extremely deeply buried coins. Many of the VLF/TR discriminating metal detectors are so versatile that they can be used in many other treasure hunting situations, including prospecting, cache hunting, and relic hunting.

and areas that have been previously hunted should choose detectors that produce GREAT DEPTH to recover the deeper and possibly more valuable coins that have been missed. Trashy, junk-littered areas will best be searched with a type of detector that has discriminating capabilities. The amount of soil mineralization should always be taken into consideration. Some detectors will operate perfectly in highly mineralized area; some, only adequately; and some will not perform at all. The ability to penetrate mineralized soil with ease of operation should always be one of the FIRST considerations when selecting a detector for coin hunting.

RECOMMENDED DIGGING TOOLS

Many different types of tools can be used to recover coins speedily. A *combination rake and hoe* is made from a small garden scratcher. Attach a longer, sturdier handle on any type of small rake and you have the equivalent of this tool. It is meant to be used ONLY where slight digging or scratching does not damage or destroy the ground surface. It is a very popular tool for searching around old homesteads and isolated areas. With it you can quickly dig a small coin or metallic object to the

Twelve-hundred contestants line up around the field where they will soon search for more than $100,000 in prizes and money at the First International Championship Treasure Hunt sponsored by The International Treasure Hunting Society. For information on the Society and its activities, write ITHS, P. O. Box 3007, Garland, Texas 75041 or call 214-271-0800.

surface without bending over. By using the end of the rake to scratch through loose soil it is possible to expose small coins rapidly, with little effort, and without damaging them.

Other similar tools are also very popular in coin hunting contests. When speed is of the essence in competition meets, any small compact tool that enables you to expose a coin without backbending is definitely an asset. It would be wise to keep in mind, however, that most hunt officials limit the size of a digging tool in competition meets. This is only fair and equitable to all entrants. Size is often limited to approximately one-half-inch width to prevent destruction of hunt areas. (Most clubs gain permission to use parks, fair grounds and similar public ground for their events.) Also, it is mandatory that all holes be filled. Common courtesy and respect for other people's property have helped the coin hunting hobby grow throughout the nation.

The *combination screwdriver-ice pick* is probably one of the most useful and essential tools for coin hunting, especially in areas where care must be taken with turf. By careful probing with the ice pick it is possible to locate the coin, insert the screwdriver carefully under it, and lift it out. You will leave no damage to the turf nor evidence that you removed the coin. Always use a digging tool similar to this when conducting a search in parks and on lawns. If you do, you will always find the welcome mat out.

The *hunting knife* is one of the more commonly used digging tools. Care must be taken when using a knife for plugging in dry weather or the plug will dry out, leaving a brown spot. If you do employ the plugging system, merely cut a round circle as deeply down as you think the coin is buried. Generally, you will find the coin near the bottom of the plug. If it is not there, check the hole for a deeper object or inspect the plug more closely. Check it with the detector. Be certain to replace the plug and tamp it down firmly. WARNING. National outcry against "plugging" is causing many states to place detectors off limits. To further your study of coin removal, read Charles Garrett's *Successful Coin Hunting* in which several methods of coin retrieval are explained.

George Mroczkowski, President of the Gem and Treasure Hunting Association, 2493 San Diego Ave., San Diego, California 92110, is a treasure hunter of many talents. George has covered the entire gamut of treasure hunting activities and has, perhaps, during his lifetime of treasure hunting, contributed more in the way of public service than any other treasure hunter. Descriptions of George's fantastic experiences in his life as a treasure hunter are woven all through his book, *Professional Treasure Hunter,* published by Ram Publishing Company.

YOU CAN BE SUCCESSFUL

If you have purchased a quality detector, gathered your digging tools, chosen a likely looking lawn where the house is relatively old, and proceeded in the normal method of searching ... YOU ARE ALMOST CERTAIN TO FIND COINS. This is why the coin hunting hobby is the largest in the world — anyone can be successful at any time, in almost any place.

If you are an avid fisherman, you do not have to become a professional treasure hunter just to search for coins. Simply purchase the detector of your choice and carry it with you each time you enjoy your primary outdoor sport. Then use the detector to find valuable coins when the fish aren't biting! The same idea holds for the businessman, hunter, snowmobiler, backpacker, cowboy, and golfer. Just think of good coin hunting places you have seen. Think of the hours of interest a metal/mineral detector will create for your friends and family. When you have an extra hour to fill, remember that all over the world coin hunting produces more excitement and fun than many other hobbies.

For those who wish to pursue the hobby to its fullest, we again recommend *Successful Coin Hunting*, published by Ram Publishing Company. It covers all phases of coin hunting with all types of detectors and lists hundreds of interesting places to hunt. It is the most informative book of its kind; no coin hunter can afford to be without it.

CHAPTER 13

Searching the Ghost Towns

All over the world, people enjoy searching ghost towns or "ghost-towning," as hobbyists call it. Finding jewelry, old coins, a buried treasure, relics or antiques dating back to before Christ, and many other items of yesterday can be considered ghost-towning. Some items will have no value except for historical interest, but that alone is enough to satisfy many treasure hunters and keep them searching.

Wherever people have gathered, camped, or lived will produce collectibles. There are thousands of abandoned town sites, old forts, homesteads, farm house locations... the list is endless. Wherever people have been, old coins and nice relics can be found. Finding a place to search will never be your problem, just the time needed to pursue and enjoy your hobby. Remember that most of the surface items were picked up by relic and antique hunters long ago. You will need a good all-purpose metal detector because most objects will be concealed or lie below ground surface.

Almost any type of detector will allow reasonable success. The TR, VLF, BFO, and PI types will, to some degree, find relics and coins. The PRG will operate perfectly in non-mineralized areas and give good identification readout on targets. The PI and VLF types will not be affected by the ground minerals. These two types are super-sensitive and would miss none of the deep targets. If you wish valuable relics, you must expect to dig all targets. Many tin cans containing money are found around these old locations. For ghost-towning and archaeological purposes, the VLF types really come into their own. In archaeological work, you want to locate and save ALL tiny pieces of metal, both ferrous and non-ferrous. Locating them before excavation could save a priceless artifact from damage during digging and removal.

You can use TR's successfully if ground mineralization does not present a problem. Locations such as old cabin sites in remote mountainous or farming areas may have heavy vegetation and overgrowth. When this occurs in a mineralized area and the TR must be used, the instrument must be tuned in the NULL zone in order to eliminate erratic operation, in which case there will be a decided loss in sensitivity which may be acceptable. However, if you have no other type detector, go ahead and search with the TR. You will achieve some degree of success.

The BFO type detector operates with continuous sound and does not have to be scrubbed on the ground. Thus, the BFO presents no operation problem for ghost-towning. Nevertheless, all locations, whether desert or mountain region, are different. You must take this fact into consideration. The BFO does not, however, have great depth and is affected by ground minerals.

Relic hunting does not call for the use of a discriminating circuit because a ghost-towner will want to recover ALL metallic items. The average detector, therefore, will perform quite well, whether or not it will discriminate.

If your goal is to detect small, coin sized objects and at the same time be able to gain enough depth to find shallow relics, the standard 7½ or 8-inch searchcoil is the most logical choice. It is small enough to detect old coins but large enough also to detect larger relics at reasonable depths. If your choice of detectors is the BFO type, you might consider the independently operated dual coil. One of the duals manufactured features a 6½-inch coil and 3½-inch coil. The small coil enables you to pinpoint the smaller coins. Smaller searchcoils such as those used for coinshooting can detect coins but have less depth than the middle-sized (12-inch) loops, especially when detecting larger objects. The larger searchcoils used to search for deeper treasure and money caches would locate the relics but fail to detect small single coins. If ghost-towning is intended to be your sole hobby, you might consider the purchase of a high-quality detector with the middle-sized, 7½-inch to 8-inch searchcoil. If you wish to extend your activities to look for deeper caches or to prospect, choose a VLF/TR detector. No matter what type detector you own or consider buying, the ghost towns will provide an endless variety of interesting metallic objects that will contribute greatly to the enjoyment of this fascinating hobby you can pursue almost anywhere.

Now, with your detector chosen, tuned, and ready to go, all you have to do is locate a site that may have accommodated a number of people sometime in the past. With your detector, spot-check several areas to see if anything is there. It does not take much to confirm a possibly productive location — perhaps an old nail or some other meaningful object — and doing so will insure that you are in the right area. If you have sufficient time, sweep the entire area with your searchcoil. If time is limited, choose the most likely looking spots and pay close attention to all signals. Sometimes an old relic will deteriorate to the point where it may not produce a pronounced signal.

Your detector dealer should stock several books on the subject of where to search. All will help to familiarize you with different areas. It would take volumes merely to touch the surface of this exciting hobby, but you will find that common sense and a good detector will generally take care of all your problems. Few, if any, ghost-towners return empty-handed. Perhaps your vacations take you to many unusual and interesting areas, possibly ghost towns and abandoned mining areas. The relics and old coins are there just waiting to be found.

Whenever the opportunity presents itself, locate barbed wire with your detector. Any type of detector can be used to locate buried coiled or single strands of all types of wire. The detector response may be either positive or negative, depending upon the type of detector you are using, the type of wire, how long it has been buried, and whether it is a single strand or coiled. It is quite easy to trace buried single strands.

Next to your detector dealer or manufacturer, the best source of information is a library. Its use costs you nothing, and in a few hours a lead can be quickly researched. This is possible even when on vacation or traveling. The librarian will guide you in selecting reference books and maps. Countless thousands of people on vacation travel through the most interesting areas without even stopping! If they do stop they generally receive little help or information from questions put to the local inhabitants and become discouraged. Remember, the next time you wish information regarding ANY area go to the local library and ASK. This is the source used by most professional treasure hunters. There is no end of places to search.

Leave an area as you found it — clean — and be sure to refill all holes. Use courtesy at all times. You might want to go back and dig that last signal you got just when you decided it was getting dark and you had to hike back out to your car.

BUILDING AND CABIN SEARCHING

It is easy to search buildings, most of them, anyway. Any amateur or beginning treasure hunter can tune a metal detector inside a building constructed of wood or other non-conductive material and tell instantly where coins, a can, or a box may be hidden in a wall.

Practically all types of detectors will work perfectly under these circumstances, provided the structure does not have nails and other small ferrous targets present. Many dealers will demonstrate their detectors inside their shops, perhaps on a wooden floor. However, this procedure can be somewhat misleading to the prospective buyer because no mineral is present,

Mr. B.A. Stevens holds this large, heavy slug of melted silver coins. Frequently, when buildings burn, many caches become melted gold and silver globs of metal. VLF/TR metal/mineral detectors with continuously adjustable discriminating controls can be set to reject various large objects such as tin cans, while accepting large pieces of precious metals.

such as there may be in an actual field operation. Some beginners may also be confused as to which type of detector operates best when used among nails or other small ferrous junk. Since most dwellings, even hundred-year-old log cabins, have some ferrous material present, correct selection of an instrument is all important.

Generally, a cache that is still intact will be in some type of metal container. You will want to detect all metal objects and, if possible, try to evaluate their sizes. Such evaluation can be very difficult if nails and other small conductive items are imbedded in the walls, ceilings and floors. Different detector types react differently to small ferrous items, not only to the type of item but also in relation to the way in which the item faces the detector searchcoil. The following paragraphs provide a reasonable description of the way each type will react under average conditions, in average surroundings.

Ghost-towning is an extremely profitable, often rewarding activity. Many caches, including gold nugget and money hoards, have been found around old cabins like this one. Here, a ghost-towner searches for an old prospector's cache that is supposed to be hidden somewhere near by in this ghost town of Bourne, Oregon.

PRG TYPE

The sensitivity, stability, and target identification of the PRG is equal to or better than that of any detector currently on the open market. The PRG type detector is heavy (almost seven pounds) and may prevent efficient overhead or tight space maneuverability. It may be too expensive to purchase to use for a relatively simple operation such as building searching. The PRG performs almost perfectly on or near salt water beaches relatively barren of mineral content, and perhaps it should be reserved for use where it gives excellent performance.

PI TYPE

PI detectors are super-sensitive to small ferrous targets (such as nails). Since it does not employ an additional circuit for discriminating purposes, it would be useless where these small ferrous targets were in abundance.

Use your PI where it performs best . . . in security matters and searches conducted by trained personnel, plus beachcombing for coins where ferrous trash is not abundant.

TR TYPE

Standard or conventional high-frequency TR's are practical for building searching under certain circumstances.

Small ferrous targets will produce different indications, depending on their position in relation to the searchcoil. A nail pointing toward the searchcoil may produce either a positive or negative response, depending on the configuration of the searchcoil. The TR type is a quick response instrument, almost always responding wide open (full audio signal) so there is no way to determine the size of the target. The TR leaves much to be desired when you are attempting to read, among all the nails, what is behind a log or wall. The response of any type TR is simply too quick to be practical for building searching and requires the operator to investigate all tiny indications. If it is a have-to case, it could be used for building searching. There will always be, however, much room for doubt as to whether you missed a small cache among all the other small indications, especially if you must detune the TR for use among the other metal. It simply does not perform efficiently where detuning is mandatory for ease of operation.

BFO TYPE

BFO's do not see most of the nails used in wall construction, especially small-headed finishing nails. Even when used among other metal the BFO will continue to gain in beats, enabling the operator to gauge the size and distance of individual targets. The BFO operates with a continuous beating or humming tone which prevents missing even the slightest indication that might result from among the rubble of small ferrous material. The BFO is the slowest response detector among all the types, but this characteristic can be very advantageous under certain circumstances. With experience and patience it is possible to gauge the size and depth of a target.

VLF TYPE

The VLF is the most practical, certainly the most sensitive, of all detectors in the TR family of instruments. Let's say you are looking for iron (money) containers. If your VLF detector does not have a discriminating circuit, you may have problems. The same would be true of the PI type detector without

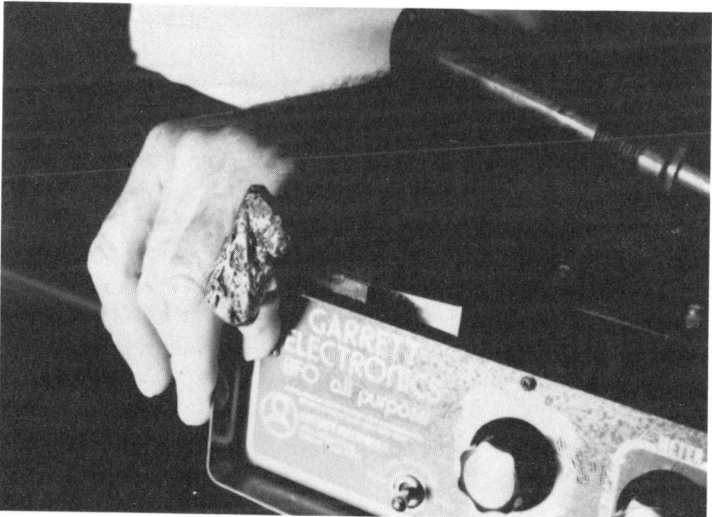

This two ounce gold nugget which has been made into a ring was found by a treasure hunter who was searching in a ghost town.

Charles Garrett uses an underwater detector to search for shipwreck site coins and relics. The sea floor in certain locations is literally covered with interesting shipwreck artifacts, some quite valuable; it is one of the most nearly untouched areas left in the world. By using the right detection equipment, archaeological teams can improve the efficiency of their locating and mapping of important underwater recovery sites.

twin circuit capability. If your instrument is one of the new VLF/TR types with variable discrimination, you have a tremendously productive detector at your fingertips.

Switch the circuit into the TR discriminating mode, adjust the discrimination control to zero, use a large nail for testing, and increase the discrimination control until the nail is simply ignored. Advance the tuning control toward "metal" until the audio is similar to a BFO constant tone. You may now freely move the searchcoil approximately four to six inches distance from the nails present in rafters and ceiling joists and the audio response will stay fairly even, neither decreasing too much nor increasing. This type of adjustment will enable you to locate all ferrous objects larger than nails. If the sound is still too erratic or jumpy, DECREASE the sensitivity control to allow for easier operation.

Too much discrimination can cause the detector to reject a tobacco can, small iron box, etc. Too LITTLE discrimination

The iron mineral content of the ground is extremely heavy in this northwestern United States ghost town location. Areas like this one produce many good relics.

allows the detector to response wide open on small nails, and you may not be able to distinguish increases produced by the larger ferrous targets. Almost regardless of the amount of increased discrimination, the VLF/TR will still respond to non-ferrous targets as small as a coin. This characteristic is one noteworthy advantage of the VLF/TR types over other types, especially when used in building searches. A small amount of practice in your home will convince you of the capabilities of these instruments.

NON-FERROUS CONTAINER CACHES

Searching for coins in glass jars thought to be concealed among large pieces of junk iron can be accomplished very easily.

The procedure is different from the search for the large iron containers. You must conduct your search in the VLF (non-discriminating) mode and then identify it by flipping to the TR discriminate mode. Follow this procedure.

Ah! A nice fourteen-karat Masonic ring found at a depth of seven inches is added to an already bulging pocket of finds that were made on this day. Bob Tatum has an impressive collection of coins, rings, relics, and other valuable articles that he has found during his many treasure hunting field trips. Thanks to modern-day technology, treasure hunters everywhere can enjoy the rewarding pursuit of treasure hunting.

Turn your VLF/TR on, adjust the tuning control for slight audio, switch into the TR mode, adjust the discrimination control until the detector will reject large ferrous targets (pipes, tin cans, etc.). Switch the detector back into the VLF mode and start your search. Pinpoint each target (you can use the detuning method). Move the searchcoil slightly to one side (off target). Switch into the TR discriminating mode and move the searchcoil back over the target. If the target is non-ferrous (coins, brass, etc.), the unit will respond positive. If it is an iron target, the audio will decrease. These instructions are outlined primarily for wall searching and the searchcoil should be pressed flat against the wall during the discriminating procedure. Practice by locating a known iron target concealed in a wall. Position a large piece of brass, aluminum, or bag of coins near the iron target. Practice by pinpointing the iron target and then identifying it. Next, pinpoint the good target in VLF mode, switch to the TR discriminating mode, press the retune button, and the detector will respond positive, even though it is only a few inches from the iron target. Care should be taken to allow the searchcoil to follow the wall in direct contact with the surface when attempting to identify all targets.

The preceding paragraphs discuss only a few of the feats the VLF/TR type detectors will perform. Several attempts may be necessary if you are not familiar with the retuning system so it is best to practice. However, you will quickly become sure of yourself and the detector and soon you will discover many more enlightening search methods.

VLF/TR detectors that will not reject large tin cans while still accepting small coins will not produce these results. However, most quality VLF types will do this successfully. VLF/TR types that do not have the variable or continuous discrimination control adjustment are also incapable of achieving these necessary settings. If your instrument does not have a variable discriminating control, perhaps you can have one installed at small cost.

NOTE: When you slide the searchcoil over the wall surface to pinpoint targets, the searchcoil will sometimes come very close to or in contact with a nail. A small sharp response will be heard. You should soon be able to quickly identify these small targets. Either ignore them or pull the searchcoil away from the wall an inch or so to double check.

During the early years, when VLF types were first made available to the public, many operators became disgusted when they were unable to operate the detectors among metallic targets. With the introduction of the instant retuning method, this problem was eliminated and now, even in spite of their supersensitive response, they are more productive than less sensitive detectors. Remember the saying, "You can turn a highly sensitive detector down, but you can't turn a low sensitivity detector up!"

LOCATING ANTIQUE BOTTLE DUMPS

Many ghost towns and other areas that have produced good bottles are constantly being reworked by collectors using metal detectors. Locations that have been searched time after time are still producing. As the average collector gains knowledge from books on bottle collecting and from information on new detectors, he is becoming aware of the fact that he just may have left some real goodies. The metal detector will never replace the shovel and bottle rake, but it certainly can find concealed dumps others have missed. Since the demand for rare, embossed whiskey bottles remains high, consider the worked-out areas that previously produced good finds. Common antique bottles have dropped somewhat in price, but the rare ones have increased at an unbelievable rate. This has always been true of priceless antiques, regardless of depressions or inflation. Never dismiss or leave a good location just because it has evidence of prior digging and previous searches. It is almost impossible for any hunter to clean a site completely. Consider well the advantage of using metal detectors to aid you in your search.

VLF/TR's equipped with large searchcoils will produce the most depth. Some difficulty may be experienced with extremely small trash targets, but this sometimes leads you straight to the hidden dump site. As mentioned several times previously, failure to dig and investigate targets may result in missing valuable relics and bottles. It should be remembered that many VLF/TR's cannot utilize large searchcoils and are incapable of producing enough depth to locate deep dumps.

Ghost-towning and bottle hunting are popular activities enjoyed throughout the world. Even though general interest in bottle hunting has declined somewhat, the interest in rare bottles continues and their value has steadily increased. Australian Keith Jenkins who owns a treasure shop at 67 Hannan St. in Kalgoorlie, Western Australia, digs a few good bottles he found in this bottle dump in a nearly deserted ghost town. He located the bottle dump while he was searching for nuggets. He had to chase this partially tame but hungry kangaroo away from his digging area!

Generally, any type detector can be used, but you *should* use the larger searchcoils to gain extra depth. A few old dumps may be relatively shallow, and the standard coils will be adequate, but most of the *really old dumps* are rather deep. In this case, larger searchcoils are necessary (the larger, the better) as the dump is usually a mass of metal and, too, the speed with which you can cover unknown search areas will be of prime importance. You should select shielded searchcoils. These coils eliminate the interference from grass and weeds that is usually encountered around bottle dumps.

You will find a ground probe helpful. When a signal is received, insert the probe before you dig to try to define what and how deep the metal is. Sometimes there will be NO glass in a small trash dump. By using the bottle probe in conjunction with the metal detector you will save endless hours of unnecessary digging.

Occasionally, you may discover a treasure cache. The large coils give fantastic depth, and you may just recover some deep items that other searchers missed with smaller coils.

CHAPTER 14

Prospecting

There are three rules to follow in your electronic search for gold. Of course, we can't guarantee instant success, but we can assure you that if you follow the electronic prospecting rules we present in this chapter you will greatly increase your chances of finding gold and other precious metals.

FIRST. Choose the correct TYPE of detector for prospecting. This does not necessarily mean some particular brand name or model.

SECOND. You must have patience. Learn to understand your detector fully and become proficient in its use.

THIRD. You must learn the correct place to start your search. No one finds gold or other precious metals where they don't exist. Stick to known, productive mining areas until you have become familiar with the telltale signs of mineral zones.

Let us now discuss the many and varied forms of electronic prospecting that are open to you. In the following paragraphs we give you the basics you will need to know in order to begin on the right foot. If you wish to study further, we recommend our book, *Electronic Prospecting*. It is complete in every way and explains every facet of prospecting with modern VLF/TR metal/mineral detectors.

METAL/MINERAL ORE SAMPLE IDENTIFICATION

METALS. Metals such as gold, silver, copper, and other valued metals are natural non-ferrous metals which will respond to any detector as *metallic*, provided they are in a conductive form and in sufficient quantity to disturb the electromagnetic field of the searchcoil. Some high grade ores are in tellurides, sulfides, and other forms, and are NOT conductive. Most free milling ores, however, that contain high grade metal in conductive form will produce good response.

MINERALS. As far as the electronic prospector is concerned, the only mineral that any type of metal detector recognizes as "mineral" is Fe_3O_4, magnetic iron oxides (in other words, magnetic iron ore or magnetic black sand). It is simple to determine if the ore contains a predominance of either metal or mineral. If the specimen of ore contains *neither* metal nor mineral, your detector would produce no indication. There is, of course, a remote possibility the specimen may contain electrically *equal* and *exact* amounts of metal and mineral. In this case one would balance out the effects of the other and no indication would be received from the detector.

If the specimen reads as "mineral" this does not mean metal is not present, only that there is a predominance of mineral. If the specimen reads as "metal" you can be certain that it contains metal in conductive form in such a quantity that you should thoroughly investigate that specimen. This factor makes the metal/mineral detector the most important tool of today's successful prospector and miner.

BENCH TESTING ORE SAMPLES

Place your detector prone on a wooden table or bench. Depending on the type of detector, tune as described below. BFO type: tune in the METAL mode of operation, adjusting the sound to a moderate beat to enable faint signal changes to be heard clearly. VLF/TR type: flip mode switch to TR mode, adjust discrimination control to ZERO (any amount of discrimination would destroy metal versus mineral identification), adjust tuning control to obtain slight audio tone. (Be sure to remove all metallic objects from your hand, i.e., rings, watches, etc.)

Quickly move an ore sample toward and away from the center of the searchcoil. Test with the center of the coil as this area will give the best response. If the sample contains neither metal nor mineral, or has electrically equal amounts of both, you receive no response. If the sample has a predominance of metal in a detectable form, you will hear a slight sound increase as the sample comes closer to the searchcoil. If you are using a detector model incorporating a sensitivity meter, you will see a positive signal or pointer movement, indicating the presence of metals. If the sound level decreases when the sample approaches the coil, the sample contains a predominance of mineral or natural magnetic iron (Fe_3O_4). This does not mean the sample contains no metal, only that the sample contains more mineral than it does metal.

A key advantage of using a metal detector to test ore samples is that it will indicate the presence of metals in conductive form. That is, you can be absolutely certain whether the specimen contains a metallic substance. The detector will not react positive unless some form of conductive metallic substance disturbs the electromagnetic field.

The best way to learn these methods is to practice. Obtain samples of galena (lead), silver, gold ore, and just plain rocks. By conducting your bench test you will become familiar with the type and amount of response to low grade and high grade ores. Also, many types of ore do not respond to metal detectors. Only those containing metal in conductive form in sufficient quantity to disturb the electromagnetic field will cause a detector to respond positively. For example, a large garnet will respond as mineral (negative) on a super-sensitive detector because the garnet contains magnetic iron. When checking samples that have responded as metal, you will generally notice a metallic appearance on the inside. When samples that appear metallic but respond as mineral are sawed into slabs, you will generally notice a streak of magnetic iron on the inside. Since there was sufficient iron to override the small amount of metal, the garnet sample responded as mineral.

ORE DUMPS AND TAILINGS

Using VLF/TR metal/mineral detectors to work yesterday's forgotten ore dumps has become a very profitable international pastime. Often, large mining companies were after only certain minerals or metals. The human eye could not see inside the ore, and many good pieces were discarded on the dump. The electronic metal detector can detect metals (of conductive form) inside almost any type of rock. Some dumps have been completely reworked for the minerals and metals that were left behind. Others are just awaiting the electronic prospector who is using a sensitive metal/mineral detector.

Near Sumpter, Oregon, in the late 1800s, a tremendous amount of prospecting and gold dredging activity took place. The wooden foundation seen in the background is the skeleton of the large dredge that once worked this area. Gold recovery was high as the giant dredge sucked up tons of sand, gravel, and huge boulders like this one and piled them onto the bank. These dredges were not 100 percent efficient, however, and many large nuggets went unnoticed into the tailings pile. The new VLF/TR detectors can be used quite profitably to scan these areas, even though the jumbled background matrix contains an uneven mineral content. Modern metal detectors do a very effective job of penetrating through the rock and mineralization to detect large nuggets.

A major reason for failure in this pursuit is that the average treasure hunter searches old mine dumps for high grade ore using the same methods as one would in usual treasure hunting. The searcher simply expects the detector to respond to small amounts of metals among a background (matrix) of extreme mineral concentration. It is not reasonable to expect any detector — no matter how sensitive — to detect small specimens of metal in a mass of rock that is generally saturated with iron or is a complete jumble of low grade metals which produces a background of metallic type response.

To search through a dump or tailing pile that may contain high grade ore, lay your detector prone on the ground. Tune it in the TR (non-discriminate or metal) mode. Select a few small samples of rock from the pile to test for metal content. If after a reasonable period of time you do not find any metallic indications, move to another area of the dump. Some areas of the dump will be more profitable than others because during the working period of the mine there was only a certain portion of the dump that could have received the tailings from the vein. The rest may be only debris from the mine's shafts and tunnels. If after prolonged testing you do not recover at least a few metallic specimens, the dump may simply be one where the ore was of a type to which a metal/mineral detector could not respond. The ore may also have been of a very low grade composition of little value.

It is wise to do your research and pick areas that produced the free-milling type of high grade ore which will respond positively to the detector. Most old-timers merely wet a rock sample to highlight the gold and then took only the high grade jewelry ore. Jewelry ore is worth much more than the gold content of the specimen. Someone else has already done the digging, and all the searcher has to do is to grade or analyze the discarded rock left on top. A good quality metal/mineral detector will enable you to do this easily, provided the metallic content is of a conductive type and rich enough to respond.

We cannot stress enough that millions of dollars lie unnoticed on small and large ore dumps, in plain sight of everyone. Anyone with reasonable ambition can use a detector to excellent advantage on these discarded rock piles. You will be surprised at the valuable specimens you recover. Remember that a small rich specimen is worth many times its precious metal content.

SEARCHING OLD MINES AND MINE FLOORS

There is no difference between old mines and new mines except that the older tunnels and shafts were generally worked by less modern methods so that there is greater possibility of finding missed pockets or hidden veins. Great care should always be taken in the exploration of any abandoned shafts or mine tunnels. Many have become unsafe over the years and the danger factor is high. It is recommended that one always search such areas accompanied by a companion.

The tunnel floor is one of the most frequently overlooked yet most productive areas in old mines. All the high grade ore had to come through the main tunnel before dumping or milling. It is almost impossible to move a large amount of rock, whether

A.M. "Van" VanFossen prepares to high grade a silver mine. He will enter this silver mine to search for native silver veins. Van has spent considerable time researching the fabulous silver mines of Mexico. If you are ever in Houston, Texas, drop by to see him at his treasure and mining equipment shop located at 2803 Old Spanish Trail.

by ore cars or by hand, without dropping a few pieces. These small ore samples were generally covered by natural debris as the workmen drilled deeper into the earth. High grade ore has lain here unnoticed for many years, just waiting for the electronic prospector who employs a quality metal/mineral detector.

Inexperienced searchers usually enter the mine and quickly sweep or search the tunnel floor with the detector, much the same as they would conduct an outside search. This method is almost always unproductive because the fact has not been considered that the mine's mineralized background distorts most responses. Use a quality VLF/TR detector to conduct a successful floor search.

To adjust the VLF/TR, switch to the TR mode and set the discrimination control to zero. This setting will correctly identify metal versus mineral. Place the detector in a prone position as in bench testing, and use your rock pick to dig under the top debris. Test small likely-looking samples. Move around to test many different spots, constantly keeping in mind that large high grade samples would have been seen by the original miner and ONLY smaller pieces would have been overlooked. However, due to the increased price of gold and the rarity of good specimens, floor searching can be a very profitable venture. Old mine floors are very productive locations for small but valuable, maybe rare, ore specimens.

If the ore is of a conductive nature, valuable specimens can be detected with a super-sensitive detector. Any unusual piece of rock should be thoroughly investigated. The metal/mineral detector will not always be foolproof in this type of searching, but it will enable you to sort unusual and out-of-the-ordinary samples from the common rock.

SEARCHING FOR MISSED MINE POCKETS AND VEINS

Pockets or small concentrations of ore occur at varying depths. Pockets and small ore bodies have often been missed

Curley Jones, Robby Robinson of Houston, and miner Manuel inspect a native silver vein that was discovered in an old worked out Spanish mine in Mexico. This vein of pure silver was one foot wide in places. Without question, veins like this are well within the reach of the VLF/TR discriminating type detectors and can be detected several feet deep into mine walls, ceilings, and floors. Think of the countless veins such as this that must lie merely beneath the surface, within reach of being discovered with today's modern metal detectors.

only by inches. Vein locating also falls under this kind of searching as a vein is just a continuous streak of ore which sometimes pinches out or completely disappears. Careful searching with a metal/mineral detector does not guarantee discovery of either a pocket or a vein. However, if you are experienced and the pocket or vein will respond to electronic detection, you have a good chance of hitting "pay dirt."

The technique of searching old mines with a detector is mostly misunderstood. Most early-day equipment and instruments were not designed to operate inside mines or in highly mineralized areas. Underground searching differs slightly from searches conducted above ground. Mine shafts and caves are often highly mineralized and require different methods and equipment. Your choice of detector types should be the VLF/TR.

VLF/TR's are the best performers underground when high mineral content (magnetic iron) is present. Turn your VLF detector on and tune it in the ground canceling mode. According to the manufacturer's instructions, proceed to adjust the ground or terrain zero control to eliminate the background

Bill Bosh, expert detector operator, rests a moment as Victor Moreland chips away at a spot where a metal detector gave a positive reading. The ability of VLF detectors to respond positively to both magnetic veins and conductive ore greatly improves the chances of success for an electronic prospector to locate rich ore bearing veins and deposits.

Javier Castellanos, Charles Garrett, A.M. VanFossen, and Curley Jones spent a month in the fabled Spanish silver mining country in the Canyon deCobre region of Mexico. When the Spaniards first came to this country, they wrote a letter to the King of Spain promising him that if he would come to Mexico and visit this area his feet would never walk upon anything but pure silver. He was told, further, that in some areas there was so much native silver on the surface of the ground that it appeared that it had just snowed. These modern-day searchers did an electronic prospecting survey which led to the finding of several silver veins and deposits. Javier is speaking with a Tarahumare Indian (on outside left of photograph), asking directions into a well hidden, remote area where research indicated the location of a long-forgotten Spanish silver mine.

response. When zeroing the ground adjust control, move the coil toward and away from the wall where you will begin your search. When this is accomplished, scan the walls and ceiling carefully, marking or taking note of all indications. The VLF/TR will respond to both conductive (metallic) and non-conductive (magnetic) ore veins when tuned in the ground canceling VLF mode. This is an advantage of the VLF/TR detector types because ore veins may contain a predominance of either conductive metal or non-conductive magnetic ores. A vein containing a predominance of metal can be identified quickly using the TR discrimination mode. However, probably ninety percent of all ore veins in the world contains a predominance of magnetic content (Fe_3O_4). Many may be rich in precious metals, but still be predominantly magnetic. To use some other type of detector and not be able to detect both conductive

This mule team carries the equipment and supplies of a month-long exploration trip into Mexico's Cobre Canyon area. Members of the International Treasure Hunting Society made the trip into this fabled silver mining region to make initial exploratory and prospecting investigations of this extremely remote, desolate, and rugged country. VLF/TR discriminating detectors are the recommended types for use in this very highly mineralized area. Electronic prospecting with metal/mineral detectors has received world-wide attention and has become a noteworthy, profitable form of treasure hunting.

and non-conductive ore types in one easy operation would be very inefficient and impractical.

When the detector responds, it becomes necessary to establish the identity of the ore. Set the TR discrimination control at the zero mark. Determine the exact center of the response and move the searchcoil to one side. If you were scanning the wall in the VLF ground canceling mode, flip to the TR discriminate mode. Maintain a steady audio response and scan back over the target area. If the sound decreases or stops suddenly, the target is predominantly magnetic iron. This is the ONLY substance that will lower the signal. If the audio remains steady or shows no inclination to decrease, the target is predominantly conductive. The variable discrimination control (one that is continuously adjustable) is now increased slowly to determine the amount of conductivity of the target. If you have previously practiced with your discrimination control, you have already determined the approximate spot where worthless PYRITE will be rejected. If you still receive a response after you passed beyond this approximate setting, it is very possible that you have discovered a rich, non-ferrous pocket or vein of ore. You may have difficulty in keeping the audio level steady as you analyze the target area. Do your best to keep the searchcoil at the correct and constant distance from the wall. Practice will improve your ability to scan at an even and constant height.

Remember, iron pyrites have a low conductivity factor. Consequently, they can be rejected by very low frequency TR discriminators while low grade non-ferrous ore will be accepted. This is one of the noteworthy advantages of the VLF/TR method of discrimination.

Often, rich ore will be associated with magnetite. As you prospect, obtain different samples and determine their content by assay. This procedure will help you to know what you are looking for and you will be better equipped to recognize valuable ore.

Deserted mines have been neglected over the years. The searching of old mines can be rather frustrating because of the erratic behavior of many detectors when used inside in tunnels or caves. Patience and experience with VLF/TR instruments will, however, produce phenomenal results. Just think of the high grade ore pockets and veins that have been missed by only inches!

DEEP VEINS

Thousands of deep veins crisscross the earth. Many are rich in precious metal; others lack any value whatever. Locating veins is one of the toughest and most unproductive phases of the prospecting hobby. Veins may respond either as metallic or mineral, and one has no way of knowing the content without an assay. Regardless of the predominant ore content, either magnetic or non-magnetic, the vein could be extremely rich.

In vein searching, the use of large coils is necessary since veins can run deep. It is possible to narrow the search area somewhat by finding surface ore called "float." Use small coils when searching for surface float. When you think you have found a likely location, change to the larger coils and work in a grid pattern. Pay close attention to all responses. *Investigation of irregular or unusual signals will sometimes lead to pay dirt.*

Regardless of the detector response, evaluate it by considering its magnitude, in which direction, and how far it runs. Within reason, you may judge the depth by the amount of response. This will not always provide an accurate gauge, but drilling or digging may be warranted if the width and length of the response area are unusual.

Adjust the tuning to achieve a slight audio sound and make wide swings with your searchcoil. The response area may be quite wide, and if you fail to cover enough area in your sweep you may not be able to determine the edge (or start and finish) of the signal. Deep veins are usually a composition of several metals and minerals; therefore, extreme caution must be observed when listening for signals as most will be very faint.

One of the best locations for *easy* vein searching is around the construction of new roads. Watch the deep pass cuts when traveling highways. Perhaps the workmen did not recognize the ore or mineralization for what it really was. A quality VLF/TR detector will quickly test these locations. It only remains for you to gather and test samples. Over the years many samples of semiprecious stones have been gathered by rockhounds from these very locations. Keep in mind, also, that logging roads may have exposed veins and stringers.

Most veins generally run in the same direction. This natural mineral matrix enables you to conduct your search in a definite pattern so you have a better chance to bisect ore veins. Veins have a habit of disappearing and then reappearing in the strangest places. One of the saddest and oft-told stories is ". . . the vein pinched out . . . we lost the vein . . . it just disappeared." When you recount all the "lost" veins that contained fabulous riches, you realize the value of the VLF/TR metal/mineral detector.

VLF/TR types equipped with large searchcoils will be one of the surest producers in the future relocation of lost veins. The ability to penetrate mineralized background permits much more efficient explorations. Modern prospectors who employ VLF/TR detectors will certainly have more than an even break. Past failures in searching for lost veins have made many professional miners wary of electronic detectors, but there is NO question that the VLF/TR's with *mineral-free operation and discrimination* will give renewed hope to the faithful and will produce wealth for the persistent. Did you ever wish you could see into mine walls just one more foot? Consider the shafts and tunnels that honeycomb areas which produced richly in earlier days. How many times the shaft or crosscut missed veins or pockets by only inches. Any prospector who fails to recognize the potential that exists may be passing up the easiest strike of his life.

FIELD SEARCHING FOR PRECIOUS METALS

Field searching has many meanings and covers many different facets of prospecting, including looking for rich float, ore chutes, chimneys, and deep veins.

As an example, take searching for high grade float . . . small pieces of ore broken off from the main ore body and carried by water across the terrain. Float usually moves with gravity downhill or downstream, depending on nature's quirks. When a piece of high grade float is found, it is best to try to determine from which watershed it came. Your first concern should be to search for nearby additional pieces to ascertain in which direction to start. Note: Keep in mind that there are many ways in which pieces of ore can be displaced by either nature or man. A miner might carry a piece of high grade ore for years and finally misplace or lose it. Sometimes the finding of such a piece of ore will result in the search for ore bodies not even in the vicinity of the find.

Almost always, you should conduct your search in gullies, creek bottoms, and on other terrain where you believe the heavy float had to stop. Logical reasoning will dictate where to search. Remember that the small pieces of rock are heavy and gravity eventually deposits them in the lowest place. Search uphill from your first find. While this isn't always correct, it is a good bet. If you're fortunate enough to find another piece, you know you are on the right track to the mother lode. Prospectors in many areas of the world use this method successfully to locate large gold nuggets and nugget deposits.

The best choice of instruments is a VLF/TR. Gold producing areas usually contain magnetic mineral. The best mode of operation is the mineral-free (VLF) mode coupled with the use

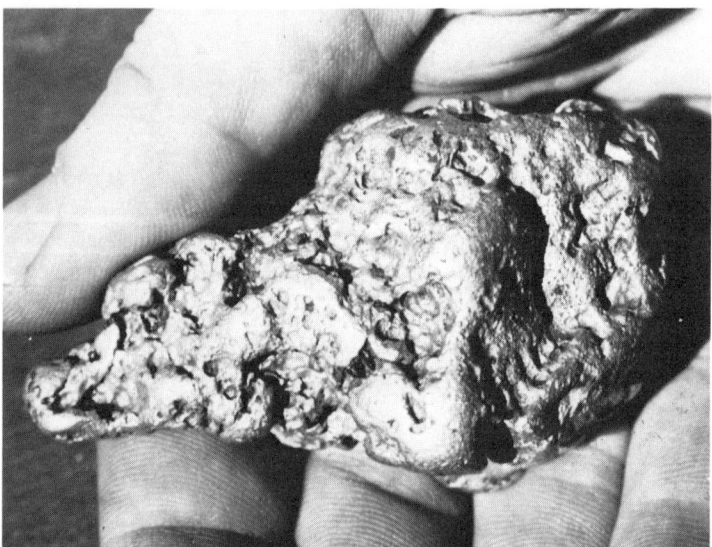

Co-author, Charles Garrett, uses a VLF/TR to search for nuggets. This location is in the Australian outback, four hundred miles east of Perth, in Western Australia. The VLF/TR's were certainly put to the ultimate test in this area where the soil contains an extremely high degree of mineralization. In certain areas in Australia (and in other parts of the world) ground mineralization is practically solid conductive iron. The nugget in hand weighs approximately two pounds. Large nuggets such as this one are found almost daily by prospectors using the new VLF/TR metal mineral detectors. The Garrett treasure collection contains many such beautiful nuggets, some weighing up to seven pounds.

of the TR discrimination mode for identifying your finds. Sweep the searchcoil in a regular pattern and attempt to maintain some sort of pattern in your search. The same crisscross pattern used in vein searching is a good bet. This always prevents one from walking only parallel to a deposit, lessening your chance of detecting it.

Save all rocks that respond as positive and are identified as metallic. Also, if you find rocks that show metal, but still respond as mineral (magnetic) bring them in for later analyzing or assaying. ALWAYS INVESTIGATE THE UNUSUAL. With luck and patience, your search may be rewarded with riches or with, at least, an interesting discovery.

ORE CHIMNEYS AND POCKETS

Ore chimneys and pockets occur where volcanic pressure has pushed material containing metals and minerals upward through a fissure or crack in the bed rock. The material cooled as it reached the surface and formed a small, isolated pocket of ore. Searching for this type of formation is called pocket searching. Over many thousands of years, the ore has gradually decomposed, and gravity has carried the decomposed ore downhill. Sometimes this ore will be in small pieces of rich float or placer gold. If the gold is concentrated enough, it has a good chance of responding to a super-sensitive detector.

POCKET HUNTING

Oregon, U.S.A., is one of the most famous pocket states. Pockets generally occur on hillsides, and gradual erosion has caused the placer gold to be carried downhill. To locate the pocket, old prospectors followed the traces uphill by patient panning. Digging down a few feet, they removed and hand-sorted the rich ore, taking only the rock with *exposed gold*. (Remember this when you search old mine dumps.) The holes left are relatively shallow and, although there is not too much rock left by which to locate the dump, there are plenty of these small mines worth high grading.

Experience has proved that the locating of rich pockets is best accomplished by using a VLF/TR detector equipped with a medium-to-large-sized coil. You need the depth of the larger coils and they will also speed your search by covering more

Roy Lagal grades jewelry gold into size and appearance categories. Much of the gold in the Garrett gold collection is converted to jewelry; the larger gold and silver specimens remain to be sold as collectors' items. The gold is gathered from all over the United States and several other countries such as Australia, Canada, and Mexico.

ground. Tune the detector in the VLF mode and sweep the searchcoil approximately four to six inches above the ground. Listen closely for rather faint indications. Signals may be faint because pockets are usually covered by a few feet of earth and the gold ore is NOT pure metal. Pockets, as a general rule, are not very large.

Often, gold occurs in porphyry. This makes digging easy as most of the pockets are relatively shallow. The gold is easy to remove by crushing the porphyry and panning the concentrates. Actually, gold deposits occur in no set pattern. It is best to remember the old saying, "Gold is where you find it."

Some prospectors and miners believe rich pockets occur in almost every mining district. Washington State, U.S.A., has probably the rarest types of gold to occur in pockets. One type is the famed crystalline wire gold that has been found around the town of Liberty, Wasington, located in the Cascade Mountains which extend from Canada to the Oregon coast. Crystalline gold nuggets are beautiful beyond expression and are found

Roy Lagal tests several types of detectors in this stream in a highly mineralized area.

only in this mountain range. This type gold is probably the most valuable ever found. Should you decide to purchase a crystalline gold specimen, consider first the story when a man once said, "If you have to ask 'how much,' then forget it; it is too expensive for you!"

NUGGET HUNTING

IN WATER. So very often the ability of metal/mineral detectors to locate small gold nuggets in mineralized stream beds has been over-advertised and oversimplified to the point where people think all they have to do is purchase a detector and head for the hills. Presto! Instant riches! Nothing could be further from the truth. Nevertheless, by following proved guidelines and paying close attention to the type of detector you choose, it is definitely possible to find the *larger* gold nuggets.

Roy Lagal instructs the miner in the use of the plastic type gold pans in conjunction with metal detectors. Here, Roy is detecting a small gold nugget that is in the pan with magnetic black sand.

Actually, many detector operators do better at this activity than dredging operators and with far less work and expense.

It is proved that search methods in water and in highly mineralized mountain streams differ greatly from the methods used on dry land. Most important, you need a detector that will operate adequately among mineralized rocks and over uneven terrain. Nugget searching under these conditions is best accomplished by the use of detectors featuring complete mineral-free operation. The VLF/TR types designed and engineered in the VERY low frequency range (three kiloHertz-plus) will best accomplish this. These detectors will penetrate magnetic black sand. Depending on the operator's skill, large gold nuggets can now be found under mineralized rocks and in black sand quite successfully. The specially designed VLF/TR types are advanced design, super-sensitive instruments and should not be confused with the popular standard coin hunting TR (IB) detectors.

Adjust the detector's VLF mode to nullify the negative effect of the highly mineralized stream bed. Operate the searchcoil about one to four inches above ground, moving it slowly

Peter Bridge, President of Hesperian Detectors, 66 Wellington St. in Perth, Australia, is Australian distributor for one of the major United States detector manufacturers. He uses a VLF/TR detector to search for nuggets in a Western Australia outback gold field. Peter and Charles Garrett have traveled throughout Australia searching for gold and testing metal detectors over every type of mineralized ground. Australia is famous not only for its large gold nuggets but for its oftentimes heavily iron mineralized gold fields which are nearly impossible to work. Peter is a well-known mineralogist who has worked for the Australian government, as well as for private industry. He is highly respected throughout Australia in mineralogical and geological circles.

over the search area. Many manufacturers provide smaller searchcoils for nugget hunting, but careful field testing will prove the seven to eight-inch size to the be the most practical, *even on small nuggets*. Earphones may be necessary if background noise (such as running water) distorts the audio response.

Keep a plastic "Gravity Trap" gold pan with you, plus a small garden trowel or shovel. When you think a metallic target has been detected, slip the shovel carefully under the spot and lift the contents into the plastic gold pan. Test with your detector to see if the detected object is in the pan. (A plastic pan is necessary because a steel pan would interfere with the detector's response.) If the metallic target is in the pan sort the gravel carefully and try visual recovery. If this is impossible try panning the contents in water. Perhaps the target is only a small piece of ferrous trash or it could be a hot rock with a different mineral (iron) content than that for which the detector was adjusted.

We know the following information is repetitive, but we realize that many of you will skip portions of this book to perhaps read only the sections that interest you most. Thus, we repeat the "hot rock" rejection procedure. Before you dig a detected target, check it by switching into the TR mode with the discrimination control set to zero. Scan over the spot and the detector will respond with the correct metal versus mineral signal, identifying the target as either a hot rock or metal. You must realize that discrimination should not be used to identify targets that might be gold nuggets. Many nuggets will respond as negative or bad because of their shape or edge contours. Simply accept the fact that if the target is identified as metal, then YOU must discern the type of metal it is. Of course, that means digging it and visually identifying it.

The metal detector/gold pan recovery method may be used on old dredge tailings. Many dredges used a small trommel screen (revolving perforated drum) to separate the rocks and nuggets. Sometimes large nuggets grizzly off (are discarded) with the rocks. Since this was a frequent occurrence, many dredge tailings may be profitably worked. Picnickers in Washington State near the old town of Liberty have found many such large nuggets among the discarded tailings. You can see some of

The silver and gold specimens shown here are from the authors' private collection. Notice the metallic appearance of these specimens. This type of native ore is highly conductive and is easily located by electronic detection systems. The specimen at the upper right was detected in the roof of an abandoned mine tunnel. The original miners overlooked several such valuable specimens in this particular mine. The other specimens are typical of ore that can be recovered from old tailing piles (ore dumps). The highgrading of abandoned mines and tailing piles is usually successful when a VLF/TR ground canceling type of metal detector is used.

these fabulous specimens displayed in many banks in that area. Any large nugget from one ounce up is easily located. Even the small five-pennyweight size presents no problem to the experienced operator who is using one of the new VLF/TR's. Of course, very small nuggets cannot be detected at any great depth.

A great amount of small metallic junk abounds in isolated mountain streams. It seems impossible that man could have been in these places in such numbers and lost or discarded so many things! As with gold, small metallic objects are heavy and eventually wind up in a streambed, carried there either by rains or gravity. Most, however, have been thrown into the water. A great many detector types are advertised as "nugget shooters" and this fact often confuses the buyer. However, only VLF/TR types get the job done with the best of efficiency. The very low frequency TR discrimination circuit is needed for good results. The discriminating circuit rejects the many hot rocks that will be encountered. Remember, as mentioned in Parts 1 and 2, the VLF/TR types must have perfectly calibrated TR discrimination controls or metal versus mineral identification may not be possible.

Since the recent introduction of the VLF/TR, many tales of success are reported almost daily from all over the world. In years to come, nugget shooting is certain to become one of the most productive forms of prospecting. The cost of quality detectors becomes unimportant when you consider that one good small nugget will pay for the instrument. Just remember, you will not get apples unless you get into an apple orchard! The same is true of gold. You must get out and look for it where it has been or might be found and use the right *type* equipment.

DRY PLACER DIGGIN'S AND DRY WASHES. The searching for nuggets in placer diggin's and in the bottom of dry washes is probably the most productive and rewarding phase of prospecting. In remote desert areas where water has never been available and where the only method of recovery was dry panning or dry washing, there are untold millions-of-dollars-worth of large and small nuggets lying just on or under the surface. They rarely are detectable by eyesight, but they lie often at very shallow depths. Investigation of low-lying areas with a VLF detector can be very rewarding. There are knowledgeable nugget hunters who have done quite well by working dry or desert areas in highly mineralized locations. Stories have been published over the years about authenticated finds of small wheat-grain size to exceptionally large specimens, and the use of the VLF/TR detector is the only practical method of locating such deposits. It is certainly one of the fastest.

The recent gold rush in Australia is a case in point. Prospectors using electronic detection methods are recovering nuggets weighting several pounds. The nuggets are undetectable by any means other than electronic prospecting with the new VLF/TR detectors.

One fascinating aspect of nugget hunting with detectors is that with each search sweep you cover much more ground than a man could shovel in the same length of time, even if you obtain only surface readings or perhaps a depth of only one inch. All things considered, you will be surprised at the actual amount of ground you can cover. This is the tremendous advantage of the use of the metal detector for prospecting. The use of the plastic "Gravity Trap" gold pan is almost mandatory here, also. You need a plastic gold pan capable of panning dry materials. The "Gravity Trap" pan uses the same type riffles as the standard dry washer. Follow the recovery instructions included with the pan for successful dry panning.

To search with the VLF/TR, simply adjust the ground or terrain control and operate with the searchcoil one to two inches above the ground. This will gain all depth possible on the smaller nuggets because of no interference from the mineralized soil. If the area is heavily mineralized, the VLF/TR types will produce the best results. Searchcoils should be in the seven to eight-inch size to facilitate ease of movement among boulders or rocks. Larger searchcoils, however, are very popular in several of the foreign countries where many nuggets, large and small, are plentiful. Only metallic objects of sufficient conductivity and a few out-of-place mineralized rocks will be detected.

We cannot overstress the following point: all detectors are not suitable for prospecting. The poor results obtained with some detectors have simply caused many an individual to quit this interesting and profitable hobby. Careful attention to selec-

One of the authors, Roy Lagal, searches for gold nuggets along a Montana mountain stream. Oftentimes, man-made objects such as nuts, bolts, and small pieces of iron are detected. When the VLF/TR responds with a metallic signal, the easy way to retrieve the object is to shovel out a portion of the sand and gravel where the detector pinpointed the object. Place that sand and gravel in a plastic gold pan and then quickly scan the pan with a detector to determine if the metallic object has been placed into the pan. If not, scoop down further, place additional sand and gravel into the pan, and repeat the scanning process.

tion of a detector will enable you to recover many items of interest from the desert or dry areas.

Research will prove quite helpful, and careful decisions as to search area can be very rewarding. Bonanza or mother lode areas that have produced unusually large nuggets or high grade ore (free milling type) are almost certain areas of success. Regardless of where you live — north, south, east, or west — gold has been and continues to be found all over the world. Again, gold is where you find it. At today's inflationary prices it takes only a small amount to add up to a nice nest egg. The system of electronic detection is bound to produce great riches for many individuals over the next few years. Dry placer areas were almost completely untouched by the old miners, and the cost of a good detector is nothing in comparison to the value of nuggets you might find.

BLACK SAND POCKET LOCATING

Black sand does not necessarily contain gold, but magnetic sand and the gold are both heavy and they tend to become

Frank Duval searches for black sand deposits with this VLF/TR metal/mineral detector while Charles Garrett pans the deposits for traces of gold. This search is being conducted in a gold district in Washington State, U.S.A.

trapped in nature's natural sluice boxes. When you are operating your sluice box, notice that the black sand also traps behind the riffles the same as in nature's sluice boxes, the rocks and crevices in running streams. Locating black sand pockets with a VLF/TR metal/mineral detector is possible both under water and on dry land. Some so-called black sand is actually grey sand or hematite. You may use a magnet to pick up or sort out the magnetic black sand, but it will not attract non-magnetic grey sand. You can heat grey sand to 2750° F. after which the hematite will become magnetic and can be picked up easily with a magnet.

Black sand pockets located with your detector should be investigated by panning. If water is available, this is no problem. If there is no water, use dry panning methods as described with the "Gravity Trap" gold pan. Do not attempt to dry pan the concentrates down too far. Simply reduce the contents down to the heavier concentrates (approximately a handful) and retain for further investigation where there is water. Often, gold

In a good producing gold area of the western United States, the authors combine metal detecting and gold panning in their search for gold. Black sand deposits can be located with metal detectors and then panned to determine if gold is present in the black sand. It is possible, but very difficult, to locate gold nuggets in jumbo rock piles such as this one. The nuggets have to be quite large.

Allan Cannon of Pomeroy, Washington, and Tommy T. Long of Boise, Idaho, search for gold in a location in the state of Washington. They used the dredge to clean out black sand deposits and then panned the black sand to recover the fine gold. The black sand deposits were located with a VLF/TR metal/mineral detector.

flakes or small nuggets will be visible during this dry concentrating procedure.

CHAPTER 15

Rocks, Gems, and Minerals

A most important and useful tool of the rockhound (besides his faithful rock hammer and patience) is the VLF/TR metal/mineral detector. Properly used it can be very rewarding, *but* it should not be used as the ultimate answer to the positive identification of all minerals and gems. Nothing will ever replace knowledge gained from experience in the identification of semiprecious stones and gems. The metal/mineral detector should be used as an added accessory to the rockhound's field equipment. It will aid in locating many conductive metallic specimens the human eye cannot distinguish or identify. Many high grade specimens of different ores can be overlooked on any given field trip. While the human eye cannot see inside an ore specimen, a good quality VLF/TR detector can.

Gold, silver, and copper are detectable metals. "Minerals" refers only to mineral which responds to a metal/mineral detector. Detectable minerals are primarily magnetic iron and iron oxide. What it all boils down to is: if the VLF/TR detector responds as "metal," bring the target in for it contains conductive metal in some form. If the target is identified as "mineral," this indicates only that the specimen contains *more* mineral than it does metal in any detectable form that would react to the detector. This means that for a few minutes' work you just might wind up with a high grade metallic sample that has been passed over for years by your fellow rockhounds.

Practice to familiarize yourself with the responses produced by both metal and mineral. Use specimens with which you are already familiar for this will aid greatly in your future testing of samples. Refer to the bench tests described for the identification of ore samples. When conducting your field search, use the detector as an *aid*, not as a complete searching tool. In other words, test any likely-appearing rocks. This kind of testing and investigation will vastly increase your knowledge and may produce valuable specimens.

In the identification of metallic ores, you will find that the very low frequency TR discriminating circuit will produce the best results. (NEVER use the VLF mineral-free mode for identification of any type of ore.) That statement is in no way meant to downgrade other detector types and models. It is only that the VLF's produce good results. Always, in the identification of metal versus mineral, set the TR discrimination to zero. This setting will reject magnetite. If you get a positive audio reading advance the discrimination to determine whether the content is ferrous or non-ferrous precious metals (silver, gold, etc.).

Many rockhounds have turned from the other types of metal detectors, primarily because of their unreliable identification of known ore specimens. This is not a fault of either the operator or detector, just erroneous selection of the type of detector. The newly developed VLF/TR types will produce correct results if you follow operating procedures carefully.

Always pay close attention to mine tailings when searching for gems. There could be high grade ore specimens and the metal/mineral detector, used as your extra eyes, will identify them. Experiment! A whole new world will open up when you become familiar with your detector. You will be able to spot check any promising-looking rocks, and perhaps you may find that worked out area isn't so barren after all.

Your detector dealer may have a wide selection of known, easily recognized gem or ore samples. Ask him to demonstrate various marginal samples on different types of detectors. The results of such testing will enable you to enjoy your hobby more profitably. If the dealer does not understand identification procedures with a metal/mineral detector, we suggest you recommend he read a copy of this book and the book *Electronic Prospecting*. The possibilities of success with a good metal/mineral detector should not be brushed aside. There are countless thousands who employ detectors in their search for missed or unrecognizable gem stones. Many of these contain precious metallics and are worth small fortunes at today's prices.

Ore dumps, rock piles, dry creek beds, dredge tailings, or any promising area may be searched with the VLF/TR type

Large trommels like this one have been used for many years in dredging operations to classify rocks and nuggets. Rocks and nuggets up to a certain size can pass through the holes you see in the side of the trommel. Nuggets and rocks larger than these holes cannot pass through and are consequently discarded. Some few early-day dredgers, unaware that large nuggets were present, allowed the large nuggets to go by unnoticed. These large nuggets can be detected with VLF/TR metal/mineral detectors. The authors are currently experimenting with VLF/TR detection methods to detect these large nuggets as they pass up the conveyor belt on their way to the dredge pile.

detectors. The VLF mode should be adjusted to the background or mineral content of the search area. The VLF mode will respond to any specimen of predominantly metallic content and at the same time it will respond to any rock containing a much higher content of magnetite than the surrounding matrix. With the VLF mode adjusted to the mineral content of the search area, any gem stock containing a higher amount of magnetite will respond. The specimen can then be identified by using the TR discrimination method.

Great possibilities exist in many areas, and such gem stock as thunder eggs, garnets in large concentrations, gem quality hematite, etc., may be detected. Just think of the possibilities that await the rockhound who includes the VLF detector in his or her equipment. Any time the gem stock contains a sufficiently different concentration of magnetite than the surrounding matrix or rubble in which it is located, it will respond in the VLF mode. Additionally, ALL conductive metallic specimens will respond. What remains is to identify the specimen as to predominant metal or mineral (magnetic) content.

Several rich silver specimen areas have been found recently in Mexico. The International Treasure Hunting Society has also reported other fantastic success stories from several members who are rockhounds. We highly recommend any quality VLF/TR type detector. It can become one of the most valuable tools the rockhound can employ. The cost is small and the rewards could be great.

CHAPTER 16

Gold Panning and Metal Detectors

Is the metal/mineral detector an aid to gold panning? Yes, the metal detector can definitely be of assistance in locating large nuggets of pure gold which is of a conductive nature. You will almost always have to make use of your gold pan to sort through the rubble of rock and sand to locate the small metallic object that responded to the detector. The target may be only a spent bullet or other metallic object placed by nature or man in the stream or dry wash. In any case, you will find the gold pan a quick way to settle the heavier concentrates and separate the lighter material. The use of a plastic gold pan with "Gravity Trap" features, is almost a must for this procedure. Obviously, you cannot use a metal pan when using a detector over the pan. With a plastic pan you can quickly determine if you have placed

When all the sand, rock, and gravel is panned away, the concentrates (gold and black sand) are left. Here Roy Lagal is pouring concentrates into a container for later evaluation.

Charles Garrett uses a "Gravity Trap" gold pan to search for gold nuggets in this dry wash.

The authors and internationally known Frank Mellish of London, England, examine the results of tests run on ore samples taken from a test site at a western United States mining claim. Frank has traveled throughout Europe, the United States, and Australia in his search for treasure and agrees that the crystalline gold found in this western state area is some of the most beautiful gold found in the world.

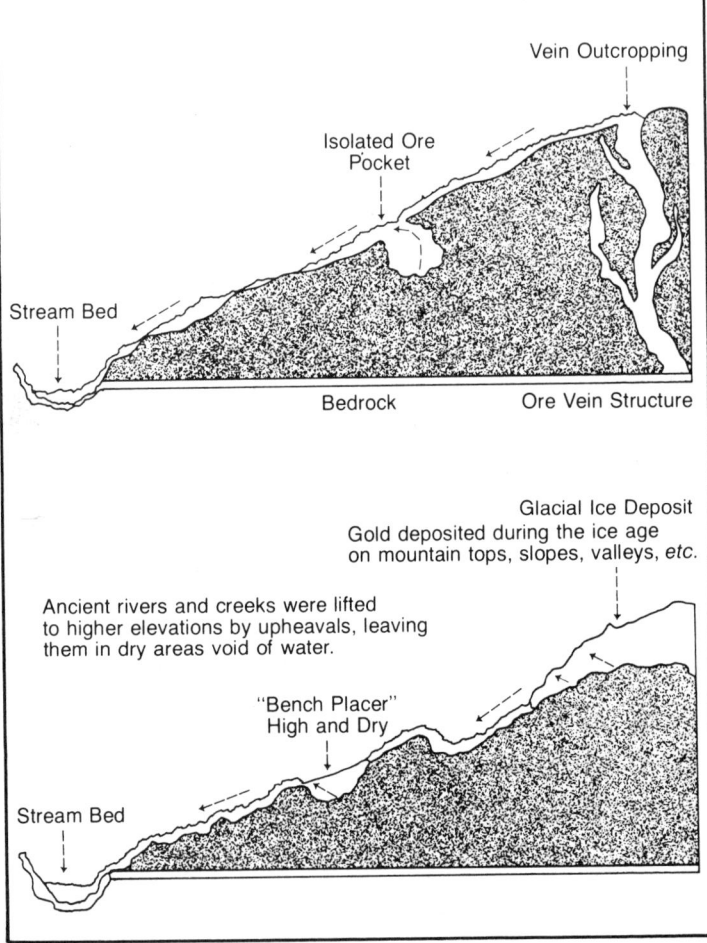

ALLUVIAL OR ANCIENT DEPOSITS IN DRY AREAS

Gold and other precious metals are released from the ore (solid state) by weathering and gravity forces, ever continuing their path to lower elevations, eventually to water immersion, and then on into the oceans of the world.

Unless the surface structure of the earth has been changed by recent upheavals, mountain slides, torrential downpours and so on, it is possible to follow the faint indications of gold to the source. Generally, this task is accomplished by patience, panning in grid patterns (made difficult by the absence of water), or visual searching. More recently accomplishment of the procedure has been made possible by newer types of electronic detection equipment. These dry areas hold great promise for the metal/mineral detector user.

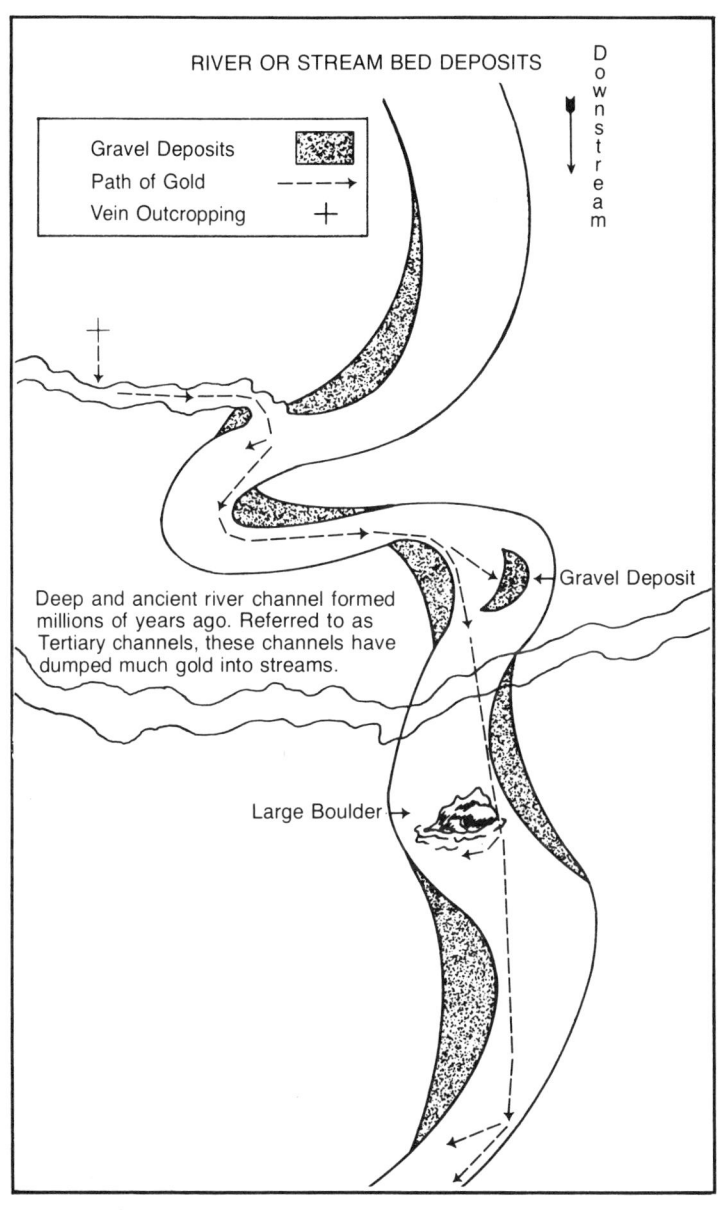

Regardless of the appearance of the present river channel and velocity of water flow, many changes may have taken place over the millions of years. River bends, current changes, amount of water flow and turbulence may have moved or displaced gold into unsuspected areas. Note the dotted lines that MAY indicate what *could* have happened. *It pays to look!*

the target into the pan on your first attempt. If not, you can dump the pan and dig more deeply or more carefully on your next attempt.

According to the *International Treasure Hunter* publication, modern-day prospectors using electronic detectors are finding large nuggets in many areas of the world. Obviously, the gold pan is not used to locate large nuggets of several pennyweight or more as they are easily visible. However, many areas produce only very small nuggets or fine gold. That kind of gold can mostly be recovered by panning.

As mentioned in previous chapters, magnetic black sand concentrations are located easily with the VLF detector. To work the concentrations, a plastic gold pan with good riffle structure is a must. Dry panning eliminates the need for carrying heavier equipment, such as a dry washer, etc. In the quest for gold and treasures, our travels have taken us to many foreign countries where older, less efficient types of pans are still being used to recover gold. Due to the international acceptance of the VLF/TR detector as an efficient tool for detecting large nuggets and alluvial deposits (provided they contain large amounts of magnetic sands), the plastic pan has been accepted as a necessary companion to the metal/mineral detector.

CHAPTER 17

Relic Hunting

War with its associated destruction has caused countless billions of artifacts to be left beneath the soil in so many battlefields the world over that treasure hunters will never be able to find them all. War has created the largest source of recoverable artifacts in existence. For hundreds, even thousands, of years, war relics have been deposited in the ground. Even today vast quantities of relics are being added to the soil in many places throughout the world.

Men and women who search battlefields for artifacts left behind by war are commonly called relic hunters. Certainly, these people are treasure hunters in the truest sense of the word. They search for artifacts for several reasons: (1), to prove history; (2), to increase their own personal collections; and (3),

Here is a close-up of a few interesting war relics. Note the extremely rare officer's pistol with built-in blade. In the center of the photograph is a piece of wood with a Minie ball still embedded in it, exactly where the ball struck that portion of the tree over one hundred years ago.

to find relics that they can sell. Many relic hunters search the battlefields for all three reasons.

Long gone are the days, however, when relic hunters could simply walk across the field and find relics with their eyes. Of course, this occasionally still happens, but some form of electronic device, namely, the metal detector, is needed today for successful relic hunting. There is no need here to go deeply into the field of relic hunting because the activity has come into its own as an extremely interesting, profitable hobby over the last few years and many articles and books have been written about this fascinating aspect of treasure hunting. Thus, in this section, we will deal mostly with the type of equipment that can be used to find war relics.

RF two-box detectors are rarely used in battlefield searching. Perhaps if a large cache of munitions or large cannon balls are sought, a relic hunter might employ one of the RF instruments. The PRG is seldom used in battlefield hunting because of its inability to operate over highly iron mineralized soil. Occasionally, a relic hunter using a pulse induction detector is seen on a battlefield. The PI instruments are easy to use, as we have explained previously, and they give very good depth in all types of soil. They are a slower scanning instrument than other types. The relic hunter can expect to do reasonably well with a pulse induction unit, but generally the PI instrument is a less desirable treasure hunting instrument, over-all. Most relic hunters have found it to be an instrument that they do not wish to use on the battlefield.

For many years BFO's were used in battlefield hunting. These detectors have found countless thousands of common and more valuable war relics. The larger searchcoils were used quite often in the search for cannon balls and other large projectiles. Today, a BFO is rarely used in battlefield searching. So many of the battlefields have been worked over that BFO's can no longer produce good quality relics.

For many years standard TR's were popular on battlefields, and even today you see quite a few being used to find war relics. Generally, the larger the searchcoil, the better. The older style 2D searchcoils were very popular on the battlefields as they had minimum iron ground mineralization pick-up. They

This is a portion of the relics displayed in the Garrett Electronics museum in Garland, Texas. These relics were found at battlefields, old homestead sites, early-day settlers' campgrounds, and Indian war battle sites. The relics were found not by professionals, especially, but by treasure hunters of both sexes and all ages, including many young people, who love to take their detectors and go in search of history. While battlefield relics are interesting to find and are often quite valuable, hundreds of reports of valuable weapons and relics found in old houses and outbuildings have come in over the years. Note the rifles and sword hanging on the wall. These items were found at various ghost town locations.

had wide scanning capabilities and gave excellent depth on most relics.

Today, however, the majority of the more serious relic hunters are using the deep seeking VLF/TR instruments with large search coils. These deep seeking instruments are producing great depth on all war relics. Individual Minie balls have been found to depths of thirty inches, truly a remarkable feat. Cannon balls have been found to depths of several feet in areas where all other types of detectors have been used day in and day out. To gain speed and enable relic hunters to search long periods of time with minimum fatigue, hip mount configurations have become extremely popular. These are accessory items that make it possible for the operator to convert a standard configuration detector into a very long-stemmed detector with good balance. Further, the operator can cover about twice as much ground in a single sweep.

Of course, even with all the modern instruments and all the attachments and accessories, there is still an important aspect left for success... the relic hunter's ability to operate his metal detector successfully. It is very important that all relic hunters develop their skill as they pursue relic hunting. Since so many of the Civil War battlefields have been searched, the majority of the shallow relics have been found. Thus, it becomes increasingly important that the relic hunter learn his instrument and learn it well so that he can pick up even the faintest of signals that indicate valuable relics lying at extreme depths. It goes without saying that the relic hunter should take great care in selecting the metal detector that will give him the most depth and do the best job. Above all, he must consider its performance capabilities rather than be led in any of his thinking by cost.

As we have said, the relic hunter would be wise to select a VLF/TR detector with the capability of using large searchcoils. Certainly, the searchcoils are much heavier than the smaller coils, but for that extra added depth so very necessary to successful relic hunting the use of large coils is mandatory.

In searching, the VLF mode should be used unless the operator is strictly interested in only non-ferrous relics. Even then, it is advisable to use the VLF mode as the standard search mode for obtaining the maximum possible depth, quickly flipping to the TR discrimination mode for target identification.

In searching in the VLF mode, the detector should be turned on and adjusted to tune out ground minerals. The searchcoil can be scanned at a height of two, three, and four or more inches. The super-depth of these instruments allows the relic hunter to hold the searchcoil as far above the ground as

While there are treasure hunters who specialize in specific facets of TH'ing, such as relic hunting, there are many who search successfully for all treasures. One such person is Pamella Wendel. She is just as at home searching for coins on the beach as she is in researching historical documents to locate fort sites. During one search of an early 1880's site, she found numerous relics, including these buttons, all of which are in remarkable condition considering the time they have been in the ground. Since there several universal type detectors on the market, all TH'ers should investigate every aspect of treasure hunting and greatly increase their shares of the rewards that come from metal detecting.

necessary to clear weeds, rocks, stumps, and so on, and still obtain excellent depth. Of course, generally speaking, the closer the searchcoil can be held to the ground, the better.

Since many battlefields are in low-lying, swampy regions, the prospective buyer of a detector for battlefield searching should be certain that the detector searchcoil that he considers buying is submersible. There may be a difference between a manufacturer's designation of splash-proof or waterproof and the term "submersible." If in doubt, ask! The detector searchcoil should be capable of being submerged at least two feet, or up to the cable connector that attaches to the control box.

It's a great feeling to search a battlefield area and find historical items. It's a good feeling also to scan a battlefield site and find relics that have been left behind, even though countless other artifact hunters have come along ahead of you! Dorian Cook, Indian John, and Charles Garrett searched the site of old Fort Pierce in Florida, and found coins, bullets, Minie balls, and other relics in areas that had been searched many, many times over the last fifteen or twenty years. Granted, most of the coins were deeper than about eight to ten inches, but, even so, finding such targets is a sign of detector operating expertise and a sure sign of improved detector instrumentation.

In June 1877, practically the entire Nez Perce Indian nation began their heroic trek to Canada. Along the way they encountered United States troops at many locations and fierce battles took place. The authors have searched most of the Nez Perce battlefield sites with various kinds of detectors. The terrain at these locations is extremely highly iron mineralized and, in most areas, extremely rocky and uneven. The only type of detector that performs well in these areas is the VLF/TR detector. These instruments can nullify the effects of the high background mineralization. The searchcoils can be operated at several inches height in order to clear rocks and uneven ground, while still producing great depth penetration. The TR discriminate mode can identify troublesome hot rocks that abound in this area. Since VLF/TR discriminating detectors have helped to overcome these extremely troublesome problems, we recommend that type of detector for all battlefield searching because, even though they may not be called upon to counter all the adverse effects of mineralization and rough terrain, they are highly capable if the need arises.

VLF/TR discriminating detectors are available with operating frequencies that run practically the entire VLF frequency band range. It has been proved many, many times over that instruments operating in the five kiloHertz range, plus or minus a few kiloHertz, produce the best results. Instruments manufactured to operate in the lower section of the VLF band generally are deeper seeking instruments. Even though the lower frequency does not permit the easiest operation in the TR discriminating mode, it does produce the best depth in the VLF mode.

CHAPTER 18

More Detector Operating Tips

CLAY DEPOSITS

Clay deposits can play havoc with a metal detector by causing your detector to give a generally rather faint metallic indication like that for a deep cache. After digging a few feet you will discover an upshoot or chimney, a deposit of beautiful clay. If you pay close attention to the condition of the clay you can tell whether it has been dug before. If it has been dug, it will have streaks of black dirt or topsoil mixed in it. It would not matter how long — two-hundred years or more — it would still show foreign dirt mixed in the clay. If you check some of the clay with your detector and receive no response, you may want to dig deeper. If the clay shows it has not been dug before, dig no further.

Clay deposits of this type are known as *neutral* clay. They contain no mineral (black sand) and no metallics. The metallic response referred to above was caused by a vacancy reading because the detector was tuned to the surrounding terrain which had a mineral, negative response on your detector. When you passed over the clay deposit, the detector simply had a chance to gain or faintly increase in volume. The same response would have occurred if you had gone over a tunnel or cave, provided it had been within detecting distance and in a mineralized area.

Of course, *positive* clay deposits may contain some type of natural conductive metallic substance and/or mineral salts that have become conductive due to the presence of moisture. The conductive nature of the wetted salts can cause the detector to respond faintly, as would a deep cache. When you test a few shovels of this material with your detector, it will respond positive, as would metal. The response will not be as strong as pure metal (like a tin can or piece of larger metal), but will be strong enough to positively identify the clay as conductive. This is not a mineral reading as most operators think but a true response on a conductive substance.

It is possible, in certain areas, that you could be getting a response from high grade gold ore or other precious metals down below or mixed in with the clay deposit. In some areas, gold deposits may be formed by a leaching process over many millions of years, and rich deposits may be found where the gold came to rest against a retaining wall, thus forming a pocket. This crystalline type gold has been discovered in mud pits and many other formations of clays, minerals, etc. The famed crystalline wire gold of the Cascade Mountains in Canada, Washington, and Oregon U.S.A., is a case in point. Prospectors who have not searched areas like this can hardly believe it when crystalline wire nuggets are found in muck and mud, high up on the mountain sides. If the clay shows a predominantly conductive content, it would be wise to have it assayed. It could possibly contain some rare earth minerals and have high market value.

TREE ROOTS, STUMPS, GREEN LOGS

Many operators pass their detector's searchcoil over a tree root, get a slight positive signal, and quickly decide the detector responded to the root. In some cases this is true, but not

Many times, sawmill operators put VLF/TR detectors to work checking for nails, spikes, and other metallic objects that are embedded in the logs. There are many operations of this type.

always. We will explain in detail what can happen and the circumstances which may cause it to happen.

Let's say you are conducting your search in mineralized (negative) ground and pass over an old dried root or place where one has decayed. You may get a slight metallic response, not from the root but from the vacancy created by the root. There was an absence of mineral in this spot which allowed the detector to gain in tone. You also get metallic responses from spots where old fence posts have rotted and left empty holes.

It is also possible to get this same metallic response over a *green* root. This kind of response is always relative to the content of the mineral in the soil. Many roots do not respond as metallic because they are not dense enough or do not have enough permeability effect to interrupt the electromagnetic field produced by the searchcoil. This is why green tree roots in certain areas with different mineralized soil content may respond either way. Perhaps you wish to test your detector on certain tree stumps or trees. You will notice that the part you are testing is completely above ground and there is no chance of a vacancy reading. You may receive a slight positive or metallic signal and, of course, you know the tree is not metal. The permeability of the dense material (combined with mineralized sap) was greater than air and the electromagnetic field was interrupted, causing the positive increase in tone. This response can only occur with BFO and TR types.

ISOLATED SALT DEPOSITS

Small salt-saturated spots may occur inland where ocean salts are absent. In neutral or negative soil a small area that has accumulated excess salts over a long period of time may become conductive when wet or damp. This situation can occur anywhere in the U.S., but you will notice only faint metallic indications when cache hunting with the larger searchcoils, depending on the water content of the soil. The faint signal may cause you to dig, thinking a deep cache is there. There is almost no way to determine the correct situation until you check the condition of soil removed from the hole. If it shows streaks and signs of having been dug before, keep on digging. You may never encounter this odd situation, but if you do you will long remember it.

MINERAL AND THE METAL DETECTOR

Some advertising on metal/mineral detectors has been known to have gotten out of hand, resulting in confusion for both experienced and beginning prospectors as to what detectors will and will not do. The confusion is understandable as there have been no reference books available. Nowhere in all the available literature can you read that NONE of the ferrous materials are magnetic, EXCEPT MAGNETITE, Fe_3O_4. Confusion has always reigned in regard to ferrous or non-ferrous, hard and soft metals. Many early-day manufacturers and advertising writers continually made reference to "ferrous, nonferrous," "hard and soft metal tuning," "responds to metal only, does not respond to mineral," "metal detector, not a mineral detector," etc.

There is no absolute necessity to require mineral tuning (negative side) on any detector merely to alert you to the presence of non-conductive iron in its natural state. The entire earth's surface contains mineral (magnetic iron oxides) to some degree. The prospector relates the mineral to that for which he is searching; the treasure hunter regards it as something in which he is not interested.

The VLF/TR type has an advantage over all other detector types because all metal in conductive form is detectable and, at the same time, so are certain mineral deposits. When commercial VLF types were first introduced, the response to out-of-place hot rocks was thought to be an insurmountable problem. Extensive research showed us how to identify these hot rocks. We also learned how to use the detector characteristic of detecting magnetic veins, conductive ore and deposits so very important to the mining industry. From the procedures given in this book, treasure hunters can identify targets as magnetic iron oxide and ignore them. The prospector using electronic equipment can detect non-ferrous (conductive) ores and magnetic veins in one efficient operation and identify them for later investigation.

When searching around big trees, there may be a once-in-a-lifetime false signal when you detect a large root. This is a rare occurrence so do not become alarmed when it happens. Your detector is performing correctly. The signal is caused simply by soil conditions combined with the mineralized sap in the root. We first noticed this effect many years ago when attempting to find nails and small metallic objects in logs. To save their saw blades many sawmill operators tried to use a standard BFO or TR detector for locating foreign objects in logs before milling. The metallic target could never be detected very deeply and sometimes was *undetectable* if it was just under the surface of the log. We called the effect "log-itis." Sawmill operators now use both the VLF and a "moving" type of detector built for pre-milling scanning.

Standard TR searchcoils can respond in two different ways on tree roots, depending on the size searchcoil you are using. Sometimes the transmitter portion happens to catch the root just right and see it as positive; sometimes the receiver portion sees it as negative. In the same way, the TR searchcoil may respond to mineral (Fe_3O_4) either way, depending on the amount of mineral, moisture content, depth, and the portion of the coil to which it comes closest. VLF and PI type detectors do not have the previously described response on green lumber. They penetrate the green log perfectly as they do not see the difference in permeability or intensity. Other than the special sawmill detectors (so-called "moving" detectors), VLF and PI types are about the only ones which can safely be employed to detect nails in green lumber.

VLF/TR TUNING SYSTEM EXPLAINED

VLF/TR detectors have been thoroughly explained in preceding chapters as to usage, advantages, etc. There is, however, almost a total lack of information or instructions available to the operator as to WHY and HOW the tuning system works. We consider the tuning as one of the most critical phases of operation of this type detector. If it is fully understood, most disadvantages can be turned into advantages.

We have discussed the two zones of tuning, metal (positive) and mineral (negative). For all practical purposes, the same principles hold true for all detector types. They have a metal zone and mineral zone with a silent area between the two zones. Many TR manufacturers simply leave off the mineral zone designation. The silent area is included in this space and the entire silent area is referred to as the null zone. The mineral zone is still there electronically, but an audible response is not produced when the tuner is rotated in that direction. On standard TR's, IB's, RF's, PI's, or PRG's the null area remains in the same place, and the threshold of entry points into audible sound remains stationary. On VLF/TR types there are TWO tuning adjustments. The control called the ground control causes the null area and the threshold point to change or shift its position when the instrument is balanced or tuned to eliminate the effects of either negative or positive mineral. Tuning control is simply a tuning device common on all detectors which is used to increase or decrease the amount of audible sound.

Here is a simple explanation and demonstration of the characteristics of ground canceling control. Obtain a twelve-inch ruler and place it on a flat surface facing you, reading inches left to right. Position a pencil at right angles on top of the ruler directly in the middle of figure six. You now have equal distance from each side of the pencil to each end of the ruler. Consider the pencil as the null area. Consider the space between the pencil and the left end of the ruler as the metal zone; the distance between the pencil and the right end of the ruler, the mineral zone. (However, remember that on the detector the mineral zone may be silent, though for practical purposes the zone is still there, even if not responding audibly.) The distance from the pencil to either end of the ruler may be considered what is usually referred to in electronics as the dynamic operating range. In other words, it represents the range or distance from where audible sound first starts (threshold) and increases to its fullest peak (saturation point).

Keep the ruler and pencil arrangement in mind and proceed to tune your VLF/TR according to the manufacturer's instructions. You will be instructed to lower the searchcoil toward the ground. If the soil is negative (mineral) and the audio tone decreases in volume, the instructions will say to advance the terrain (ground or zero) control into the INCREASE (or METAL) direction. In making the adjustment for negative (mineral) you will ALWAYS advance the control into the metal or INCREASE zone. Illustrate this procedure for yourself. Slide the pencil slightly to your left into the designated metal zone. Notice as you moved the pencil into the metal zone THE NULL ZONE (the pencil) position was moved! The is becoming more narrow; the mineral zone (on your right) becomes longer or wider. If the ground is HIGHLY mineralized the pencil (representing the null zone) may have to be moved to your left as far as numbers one or two on the ruler. Notice how this keeps decreasing the dynamic operating range of the metal zone, making the mineral side longer.

Test your VLF type detector in this same manner. After you have advanced the ground adjustment sufficiently to zero out the negative effect of highly mineralized ground, test the detector adjustment with small coins or other metallic objects. Notice the objects respond correctly as metallic. Now obtain a sample of common rock with a *high* magnetic iron content. Test it in the same manner as a coin. Depending upon how far the control was advanced into the metal zone, the mineralized rock may respond as metal.

The null zone remains stationary on both the BFO and standard TR types. In order for these detector types to detect magnetic veins, they must have a mineral zone. However, they cannot (as VLF/TR's can) accomplish *dual* detection of both magnetic veins and metallic ores while operating in the same mode. The presence of magnetite in ore veins, pockets, etc., affects the standard TR and BFO adversely, causing great loss in depth.

In VLF/TR detectors, the magnetic content of a vein or pocket enhances the detection of non-ferrous (precious metal) ores by adding to the positive response when operating in the VLF mode.

This is why we have attempted to explain in such detail the advantages of the new VLF type detectors. It is now easily possible for anyone to find conductive veins and high grade ore pockets in old abandoned mines (and working mines). The VLF type detector, properly understood and employed, will place electronic detection of precious metals forever on the map.

DETECTION OF GOLD IN MAGNETIC VEINS

Can it be done? Yes! If you are searching in a mine with your VLF tuned (adjusted) to the matrix (the inside mine wall) and cross or pass over a magnetic vein that contains a deposit of conductive metal (gold, silver, etc.), you detect the magnetic vein only as an out-of-place hot spot. (It would depend on where the ground control is set in regard to the original null zone.) If you locate the more highly mineralized area (magnetic vein) and then adjust the VLF to the magnetic content of that specific area, conducting your search following the magnetic vein, THEN you would detect the conductive deposit (gold, silver, etc.) WITH EASE. All these procedures are possible, and *highly productive* if you completely understand the VLF detector response in regard to these out-of-place hot spots that were not originally adjusted out by the terrain or zero control. Remember the ground control is continually shifting the center of tuning each time the detector is adjusted to a different mineral content. This fact must be taken into consideration when identifying a target.

KINDS OF FERROUS IRON

This subject has always created confusion among even the most experienced detector operators. Many operators believe (because they have been so instructed) that all types of ferrous iron are supposed to produce negative or mineral response. Positive responses from a target are supposed to indicate non-ferrous composition. These beliefs must be corrected.

There is no such thing as ferrous vs. non-ferrous identification, except on sophisticated sensors that recognize all types of metals. ALL TYPES OF FERROUS IRON DO NOT PRODUCE MINERAL (negative) RESPONSE. Only one type does it predominantly — Fe_3O_4, magnetic iron oxides in the natural form. For example, hematite (Fe_2O_3) in its PURE FORM is actually conductive (the same as gold). It is non-magnetic and responds to a metal detector as metallic. Hematite crystals are actually used as diamond substitutes in jewelry-making. Hematite is rated over *six* in the hardness scale and is considered a gem stock by rockhounds. It is *very* difficult to locate in the United States in its pure form (before it became heated from volcanic action). There are great hills and deposits of hematite (in red volcanic form) in the U.S., but these deposits have been hot (received heat) and will probably respond as negative on your metal detector.

What does heat do to ferrous iron? The melting point of hematite, limonite and magnetite is 2750°F. When hematite becomes hot, oxygen will be introduced into its composition and it becomes magnetic. Practically all of the hematite masses in the U.S. have been hot and are magnetic to a certain degree. They also respond faintly as mineral (negative). The reason they do not respond violently as mineral (negative), as does magnetite, is that to some degree they still retain a conductive content which helps to cancel out part of the negative response. Look at a sample of hematite that has been only slightly hot. If it is of gemstone quality it will still be black and very hard. It will have an outer coating of iron that shows the bubbles or spots where the heat changed its structure. The outer layer is all that the detector sees and the instrument will show a slight negative reaction.

All that responds as metallic on a metal detector is not necessarily metal, just *conductive*. The mineral salts (ocean salts) are a good example of this. When they become wet, they become conductive. ALL types of detectors do NOT recognize them as conductive, but the average, everyday types that we use certainly do. This is just one example of something that responds as metal that is not metal, ONLY CONDUCTIVE. Another good example is pyrite ... yes, iron PYRITE, "fool's gold." Obtain a good, high grade specimen of iron pyrite and test it on your detector. If it responds positive (metallic), look it up in a good mineral identification book. You will find it described as follows: NON-*metallic;* brass-colored; metallic luster; hardness 6 up; specific gravity about 5; melting point about 1260°F.; chemical content FeS_2; composition iron sulphide; occurrences in veins; associate mineral (metals) gold and silver; remarks-source of sulphur. (Here is another example of a ferrous substance which is conductive, the same as the other non-magnetic ferrous irons.)

After you are convinced you have a positive specimen break off a small piece, place it on a metal object, and use a propane torch to heat the pyrite to its melting point. (Propane will do this on pyrite, but on other iron and gold you would need a higher heat.) After you have heated the pyrite specimen to the red-hot point, let it cool and then test it again on your detector. It will now respond as *negative* (mineral). Crush it up and place a magnet over the crushed particles. You will notice it has become magnetic. That is exactly what happens to other ferrous iron when you add oxygen to their chemical composition. None of this information is likely to make you any the richer, but perhaps it may help explain certain occurrences and what happens to rock in volcanic actions and heat zones.

The electromagnetic field created by the searchcoil will not respond to metals in many forms, some of which are precious metal in sulfides and tellurides. It will respond only to metal in conductive form, so-called free milling ore, of sufficient conductivity to interrupt the electromagnetic field.

NEUTRAL GROUND

This term applies to any type of soil or ground that has absolutely no effect on your metal detector, positive or negative. In other words, when you tune your detector in the normal mode of metal operation and place the searchcoil on the ground, it will neither speed up nor slow down. This is called neutral ground.

Sometimes the operator will decide the ground is relatively free of mineralized iron oxides, though this is not generally the case. The earth is like a huge ball, the center of which is thought to be molten lava. Entire continents contain some degree of magnetic iron in its natural state.

Test the soil by scrubbing with a small magnet. Inspect the magnet to see if it contains any small particles of black sand. If it does, the ground contains some degree of iron mineralization, but the negative metal detector effect of the mineralization has been canceled out by some type of metallic or conductive substance. This conductive substance is usually mineral SALT from the seas which has been spread over portions of the earth. You will notice most neutral ground occurs in coastal areas. It is mostly unimportant to us why some ground is neutral. Just be thankful it is because it enables all types of detectors to operate successfully without bothersome negative soil.

POSITIVE GROUND

Positive ground will respond to the metallic mode of tuning on most detectors. That is, if you tune your unit in the normal metallic mode of operation and move the coil toward or contact the soil, you will receive a positive signal, similar to metallic responses. This does not mean the soil contains no magnetic mineral, only that the detector reads the soil as metallic. This happens because of many factors. Perhaps originally the soil had little erosion and it has received enough conductive metallic particles to override the magnetic iron present and create a positive reaction within your detector.

On some seashores there may be large quantities of black sand or magnetic iron present, but, when you operate below the tide line while the sand is still wet from conductive sea water, you will probably get a positive or metallic reaction on your detector. You might move inland where the tide has not wet the soil and the ground might respond as negative. This variance will occur depending on the amount of saturation received from conductive salts or the amount of mineralization present.

In prospecting you will discover a few mine tunnels that have enough naturally distributed low-grade metallic ore to cause the entire tunnel to read as metallic. When this occurs, the richer pockets will respond even more.

Below old mining locations in some highly mineralized areas, you will discover the entire ground reacts as metal. This

The authors relax after a hard day of field testing various types of detectors. The compilation of this book has required years of research and field work at countless sites around the world.

is usually caused by the natural distribution of fine metallic particles from low-grade ores which are of sufficient quantity to cancel or override the effects of mineralization.

VLF/TR detectors eliminate the effects of negative (magnetic) ground minerals. However, the positive effect of conductive ground may be impossible to cancel, depending on the amount of salt and moisture content. Until a more perfect system of electronic detection is attained, the VLF/TR types will remain the choice for all purpose use.

APPENDIX A

Detector Manufacturers

Bounty Hunter
1309 W. 21st St.
Tempe, AZ 85282

C&G Technology, Inc.
2515 W. Holly
Phoenix, AZ 85009

Compass
P.O. Box 366
Forest Grove, OR 97116

Fisher Research Laboratory
2130 Arthur St.
Klamath Falls, OR 97601

Garrett Electronics
2814 National Drive
Garland, TX 75041

Gold Mountain
604 Walnut Circle E.
P.O. Box 40507
Garland, TX 75040

White's Electronics
1011 Pleasant Valley Rd.
Sweet Home, OR 97386

APPENDIX B

Dredging and Prospecting Equipment

The following is a list of reputable dredging and prospecting equipment dealers. It is, by no means, a complete list, nor does the omission of a dealer's name indicate that he is not reliable.

Exanimo Establishment
Box 448
Fremont, NE 68025

Hesperian Detectors
P.O. Box 317
Victoria Park, 6100,
Western Australia

Keene Engineering
9330 Corbin Avenue
Northridge, CA 91324

Lela Harissi
106 Agiou Dimitriou St.
Thessaloniki, Greece

Canadian Treasure Trails
P.O. Box 22
Camden East, Ontario, Canada K0K 1J0

"Gold Grabber" Manufacturing Company
Professional Gold Dredges
P.O. Box 3255 (208-342-0260)
Boise, Idaho 83703

Miners Supply
Box 1301
Riggins, ID 83549

Oregon Gold Dredge
120 Monroe
Eugene, OR 97402

Treasure World
155 Robert St.
London, NWI England

APPENDIX C
Magazines and Periodicals

Coinage
17337 Ventura Blvd.
Encino, CA 91316

Dig
Found Enterprises
133 Prospect St.
Auburn, MA 01501

Exanimo Express
Exanimo Establishment
Box 448
Fremont, NE 68025

Gold Prospectors News
and Miners Journal
GPAA
P.O. Box 507
Bonsall, CA 92003

International Treasure
Hunter
International Treasure
Hunting Society
P.O. Box 3007
Garland, TX 75040

In the Steps
P.O. Box 5
Mule Creek, NM 88051

Lost Treasure
15115 So. 76th E. Avenue
Bixby, OK 74008

North South Trader
8020 New Hampshire Avenue
Langley Park, MD 20783

Prospector's Gazette
Segundo, CO 81070

The Treasure Hunter
P.O. Box 188
Midway City, CA 92655

Treasure, Treasure Found,
Treasure Search
10968 Via Frontera
San Diego, CA 92127

Western and Eastern Treasures
P.O. Box 253
Mt. Morris, IL 61054

International Club Digest's
World of Treasure
P.O. Box 7030
Compton, CA 90224

APPENDIX D
Treasure Hunting and Prospecting Schools

Gem and Treasure Hunting Association
2493 San Diego Avenue
San Diego, CA 92110

Gold Prospectors Association of America
P.O. Box 507
Bonsall, CA 92003

In the Steps
P.O. Box 5
Mule Creek, NM 88051

APPENDIX E

ARE YOU INTERESTED...

In treasure and coin hunting, relic collecting, ghost-towning, prospecting and/or nugget hunting? For free information on how to get outfitted properly and be successful in the great outdoor hobby of metal detecting visit your local equipment supplier. The new, correctly calibrated VLF/TR Ground Canceling Detectors are being used successfully all over the world. You can easily enter this profitable and exciting field.

ALABAMA: Birmingham, P & S Business Machines, 4511 5th Avenue So., 35222. (205-595-8322); **Florence,** John G. Link, 310 Colonial Drive, P.O. Box 682, 35630. (205-766-0087); **Huntsville,** Alabama Treasure Hunter, 909 Chatterson Road, 35802. (205-881-7772); **Lanett,** Belcher's Coins, 19 South 16th Street, 36863. (205-644-1881); **Mobile,** Confederate Ordnance, 2202 Government Street, P.O. Box 66075, 36606. (205-473-3731); **Oxford,** Hoffmeyer's, 1429 Snow Street, 36203. (205-831-7730).

ALASKA: Homer, R & R Detectors, P.O. Box 1707, 99603, (907-235-8200).

ARIZONA: Phoenix, Lucky Treasure World, 6005-D West Thomas, 85033, (602-247-4506); **Tempe,** The National Treasure Hunters League, 1309 West 21st Street, 85282, (602-968-9295); **Tempe,** The Treasure Shack, 2190 E. Apache, 85281, (602-968-0783); **Tucson,** Morey Detector Sales, 3825 E. Hardy Drive, 85716 (602-323-0071).

ARKANSAS: Camden, W. W. Mosley, P.O. Box 7, 768 Crestwood Rd., 71701, (501-836-5314); **Harrison,** Ozark Treasure Hunter League, Industrial Park Rd., 72601, **Little Rock,** Bill's Detectors, 5623 R Street, P.O. Box 7347, 72217, (501-666-6355); **Mountain Home,** Trammell's, 619 Baker Street, 72653, (501-425-3615); **Rogers,** L. L. Lincoln, Route 1, 158 Pyramid Drive, 72756, (501-636-6867).

CALIFORNIA: Auburn, Lo Sierra Mining Equipment, 123 Palm Avenue, 95603, (916-823-1880); **Bakersfield,** C & J Detector Sales, 3104 Pepper Tree Lane, 93309, (805-397-0641); **Bloomington,** Prospector Supplies, 868 Ironwood Avenue, 92316, (714-823-6165); **Brea,** Brea Bicycle & Sporting Goods, 141 S. Brea Blvd., 92621, (714-529-3353); **Buena Park,** Aurora Prospecting Supply, 6286 Beach Blvd., 90620, (714-521-6321); **Chowchilla,** Rencher Welding & Machine Works, 312 Calusa Avenue, 93610, (209-665-4219); **El Dorado,** Thomas Murry, P.O. Box 406, 6001 Pleasant Valley Road, 95623, (916-622-5245); **Forest Ranch,** Roy Gene Rolls, Hwy 32 at Sugar Pine, 95942, (916-342-4829); **Fresno,** Fresno Hobby & Crafts, 3026 N. Cedar, 93703, (209-226-4800); **Lafayette,** Fumble Fingers, 1027 Brown Avenue, 94549, (415-284-7406); **Lancaster,** Antelope Acres Market, Ron Farrell, 48011 90th St. West, 93534, (805-948-4190, 942-7165); **Modesto,** Gold Nugget Miner's Supply, 1302-9th Street, 95354, (209-529-5277); **N. Hollywood,** Treasure Emporium, 6507 Lankershim Blvd., 91606, (213-985-5217); **Northridge,** Keene Engineering, Inc. 9330 Corbin, 91324, (213-993-0411); **Orange,** Allied Services, 966 No. Main Street, 92667; **Riverside,** Pioneer Recoveries, 3510 Audubon Pl., 92501, (714-682-4302); **Rosemead,** Bill & Melba Dibble, 8851 E. Lansford Street, 91770, (213-287-7996); **Salinas,** B. C. Douglass, 1537 Placer Way, 93906, (408-449-1815); **San Bruno,** Dennis E. Witkowsky, Coins and Supplies, P.O. Box 772, 94066, (415-589-8179); **San Diego,** Gem & Treasure Hunting Association, 2493 San Diego Avenue, 92110, (714-297-2672), (Closed Monday & Tuesday); **San Fernando (Lakeview Terrace),** Arts & Hobbies, 12323 Forest Trail, 91342, (213-899-1997); **San Francisco,** Mining & Lapidary, 131 10th Street, 94103, (415-626-6016); **Santa Maria,** Johnny's Metal Detectors, 207 N. Broadway, 93454, (805-922-8703); **Shandon,** Price's Treasures, P.O. Box 201, 93461, (805-238-6487); **Signal Hill,** Hidden Rod Shop, 2623 Gardenia Avenue, 90806, (213-427-8060); **Simi Valley,** Gemstone Equipment, 480 E. Easy Street, 93065, (213-348-6807).

COLORADO: Colorado Springs, Terry's Treasure Hut, 1217 N. Circle Drive, Circle East Shopping Mall, 80909, (303-597-4709); **Denver,** C & D Detection Enterprises, 5885 W. 38th Avenue, 80212, (303-424-7780); **Englewood,** The Prospectors Cache, 59 W. Girard, 80110, (303-781-8787); **Montrose,** Miner's Mart, 317 East Niagara, 81401, (303-249-8752); **Pueblo,** Reg's Electronics, 4 Bonita, 81005, (303-561-3036).

CONNECTICUT: Stratford, Edward Perchaluk, 304 Circle Drive, 06497, (203-378-1660); **Suffield,** J & E Enterprises, 1242 South Street — Route 75, 06078, (203-668-0029).

FLORIDA: Fort Lauderdale, Lawson Studio, 1503 East Las Olas Blvd., 33301, (305-463-5311); **Fort Walton Beach,** James R. Ford Treasure Chest, 528 N. Eglin Pky, 32548, (904-863-1595); **Hallandale,** Silver & Gold Metal Detectors, 24 N.W. First Street, 33009, (305-457-9999); **Jacksonville,** Old Kings Road Treasure Inn, 6946 Old Kings Road So., 32217, (904-733-1928); **Lakeland,** Sight & Sound, 2024 E. Main Street, 33805, (813-644-1690); **Leesburg,** Palm Plaza Cards & Gifts, 713 N. 14th, 32748, (904-787-4661); **Maitland,** Kellyco Detector Distributors, 1443 S. Orlando Avenue, 32751, (305-645-1332); **Melbourne,** Zephyr Treasures, 2898 Zephyr Lane, 32935, (305-254-2796); **Merritt Island,** Mail Order Electronics, 200 Mustang Way 13-B, P.O. Box 1133, 32952, (305-452-8236); **Miami,** Seatech Metal Locators, 985 N.W. 95th Street, 33150, (305-693-1431); **Oakland Park,** Josh Wilson's Detector Sales, 4704 N.E. 17th Avenue, 33334, (305-776-1076); **Pensacola,** Twelfth Avenue Drugs, 2435 N. 12th Avenue, 32503, (904-433-6563); **Tampa,** Carl Anderson, Box 13441, 33611; **Tampa,** Florida Treasure Hunters, 907 23rd Avenue, 33605, (813-226-3824); **Tampa,** Treasure Shack, 3934 Britton Plaza, 33611, (813-833-9841).

GEORGIA: Atlanta, Southeastern Treasure Hunters, 985 Woodland Avenue S.E., 30316, (404-627-6019); **Decatur,** Finders Company, 225 Upland Road, 30030, (404-377-0974, Call Evenings); **East Point,** Ernest M. Andrews, Atlanta Tri-City Area, 2755 Sylvan Rd., 30344, (404-766-8141); **Waycross,** J. C. Ballentine, P.O. Box 761, Hatcher Point Mall, 31501, (912-285-3250).

IDAHO: Lewiston, Roy Lagal, Outdoor Hobby Supply, 2416½ E. Main, 83501, (208-743-1768); **Pocatello,** Powers Candy Co., Powers Home Games & Hobbies, 602 S. 1st Avenue, 83201, (208-232-1693).

ILLINOIS: Bloomington, Rene's Treasure Trove, 214 East Front Street, 61701, (309-829-4538, 829-4058); **Chebanse,** Jerry's Treasure Hunter's Supply, RR #1, Meents Lane, 60922, (815-939-3815); **Decatur,** The Base Camp, 1145 Semor Drive, 62521, (217-422-7307); **Galesburg,** Detectors Unlimited, 1671 Summit Street, 61401, (309-342-4032); **Lombard,** Electronic Exploration, 575 W. Harrison Rd., 60148, (312-620-0618); **Moline,** Hidden

Treasure, Rev. John J. Costas. 3116 11th Avenue "A", 61265. (309-797-3098); **Pekin**, Dee's Beauty Shop. 206 Reservoir Rd. , 61554. (309-346-4377); **Quincy**, Mid-West Treasure Detectors. 507 So. 8th Street. 62301. (217-223-4757); **Verona**, Gary & Karen Bennett. Indian Trail Rd. , 60479. (815-942-5290); **Waukegan**, Tom's Pool Center, Inc. , 801 North Green Bay Rd. , 60085. (312-244-4505); **Wedron**, Memory House. 1 N. Chestnut Street. 60557. (815-434-3568)

INDIANA: Decatur, O-D Western Store. Robert A. Everett. RR #5. 46733. (219-724-2097); **Fort Wayne**, A-Z Coins & Stamps. Glenbrook Center. 4201 Coldwater Rd. , 46805. (219-483-3743); **Gas City**, Phil's Enterprises. 915 E.N. H. Street. 46933. (317-674-5803); **Hammond**, J & J Coins. 7019 Calumet Avenue. 46324. (219-932-5818); **Indianapolis**, L & M Sales. 7310 Hazelwood Avenue. 46260. (317-255-4236); **Indianapolis**, The Prospectors Pouch, Indiana Treasure Hunting Headquarters. 246 S. Butler Avenue. 46219. (317-356-7343); **Seymour**, Wray's Treasure Shop. RR #5. 47274. (812-497-2537).

IOWA: Baxter, Richard Cross. 314 South Main. 50028. (515-227-3391); **Bettendorf**, Ralph Barnett. 2918 Summit Hill Ct. 52722. (319-355-6366); **Clear Lake**, Norman Treslan Construction. 4 N. 16th Street. 50428. (515-357-2255); **Indianola**, Herb Dunn Jr. Metal Detector Sales. Route 4. 50125. (515-981-4341); **Tama**, McGrew Oil Co. , 120 W. 4th Street. 52339. (515-484-2946, 489-2396); **Waverly**, Trading Post. 403 West Bremer. 50677. (319-352-9874).

KANSAS: Manhattan, Radio Shack Associate Store. 2609 Anderson Avenue. 66502. (913-539-6151); **Pratt**, Epp's Coin Shop. 112 S. Main Street. 67124. (316-672-6181, 672-6277); **Topeka**, Maxine's Treasure Sales. 5425 SW Wanamaker Road. 66604. (913-862-2872); **Wichita**, Swaim Electronics. 1430 E. Douglas. 67214. (316-262-0077).

KENTUCKY: Ashland, Gambill Locksmithing. 1004 Comanche Ct. , 41101. (606-325-7931); **Bowling Green**, Thomas D. Brenner Treasure Hunter's. P.O. Box 147. 42101. (502-781-7796); **Hopkinsville**, Urey S. Colley. 306 Fairview Drive. 42240. (502-885-2639); **Louisville**, A.F. Waller. P.O. Box 72083. 40272. (502-937-8008); **Nicholasville**, Paul Phillips. 109 Lake Street. 40356. (606-885-3648).

LOUISIANA: Benton, A-Able Treasure Electronics. 102 Duval. 71006. (318-965-0277); **Baton Rouge**, J & F Enterprises. 12211 Greenwell Springs Road. 70814. (504-272-8500); **Homer**, Hidden Treasures. 821 South Main. 71040. (318-927-6539); **Metairie**, Henry L. Montegut. 437 Aurora Avenue. 70005. (504-834-2378).

MARYLAND: Edgewater, Finders Keepers. John Reichenberg. Route 4. 3316 Oak Drive. 21037. (301-798-1833); **Glenburnie**, Frank's Detectors of Glenburnie. 408 Arbor Drive. 21061. (301-768-3157); **Westover**, Somco Machine Co. , Route 1. Box 272. 21871. (301-651-1571, 651-3964).

MASSACHUSETTS: Agawam, E & D Electronic Sales & Service. 83 Parker Street. 01001. (413-786-7190); **Auburn**, Found Enterprises. 65 Auburn Street. 01501. (617-832-3721); **Rehoboth**, Larry Violette. Box 74. 02769. (617-252-4497); **W. Springfield**, A.J. Dumais. Dumais Electronics Corporation. 37 Spring Street. 01089. (413-733-9548).

MICHIGAN: Bay City, Lloyd R. Buzzard. 1724 E. Salzburg Rd. , 48706. (517-684-4765); **Dearborn**, Huffmaster Electronics. 1537 Monroe. P.O. Box 2509. 48124. (313-278-7922, 278-1940); **Grand Rapids**, Grant's Book Store. 601 Bridge Street NW. 49504. (616-458-6580); **Lansing**, Finders Keepers Metal Detectors. 2112 Cumberland Road. 48906. (517-321-6594, 323-4250);

Union Lake, Old Prospectors Shack. 7007 Cooley Lake Road. 48085. (313-363-7328); **Wyoming**, Treasure Hunter's Supply. 3930 Burlingame SW. 49509. (616-538-1957).

MINNESOTA: Bloomington, Mid-West Metal Detectors. 8338 Pillsbury Avenue So. , 55420. (612-881-5254); **Minneapolis**, Garrett Metal Detector Specialists. 3249 Nicollet Avenue S. 55408. (612-827-3113); **Minneapolis**, Paul's Detector Sales. 4153 31st Avenue S. , 55406. (612-724-2154); **St. Paul**, Minnesota Prospectors Supply. Formerly of Red Wing. MN. 902 Goodrich. 55105. (612-226-5118).

MISSISSIPPI: Jackson, Eagle Arms Co. , 3115 Terry Rd. , 39212. (601-373-4557); **Tupelo**, Hobbies Unlimited. P.O. Box 1161. 1219 Nelle Street. 38801. (601-842-6031).

MISSOURI: Aurora, Aurora Detector Sales & Valuables Recovery Service. 303 Rock Street. 65605. (417-678-2902); **California**, Twin City Gun & CB. 500 Cooper Street. 65018. (314-796-2166); **Florissant**, The Prospector's Shack. 975 Grenoble Lane. 63033. (314-837-4703); **Fredericktown**, Allen's Hobby Shop. Court Square. 63645. (314-783-5500); **Hillsboro**, E & R Detector Sales. P.O. Box 213. 63050. (314-789-2078, 586-4263); **Independence**, Ozark Treasure Chest. 1816 Ellison. 64050. (816-252-8998); **Joplin**, Frank's Sales & Service. Route 3. Box 834. 64801. (417-781-6597); **Kansas City**, Clevengers Detector Sales. 8206 North Oak Street Trfwy. , 64118. (816-436-0697); **Miller**, Friend's Treasure Outpost. Route 1. 65707. (417-452-3852, 452-2179); **Poplar Bluff**, The Treasure Nut. 1315 North Main. 63901. (314-785-1164); **Springfield**, Radford Jewelers. 1864 South Glonstone. 65804. (417-881-7308); **St. Joseph**, Stanley Johnson Co. , 2607 So. 14th. 64503. (816-232-5163); **St. Louis**, Plateau Detector Center. 9837 Kimker. 63127. (314-842-0413).

MONTANA: Missoula, Electronic Parts. 1030 S. Avenue West. P.O. Box 2126. 59801. (406-543-3119).

NEBRASKA: Ames, Exanimo Establishment. Main Street. 68621. (402-727-9833, 721-9438); **Sprague**, L.P. Enterprises. Box 46. 68438. (402-794-5730).

NEVADA: Fallon, Scott Goodpasture. 9525 Pioneer Way. 89406. (702-867-2015); **Reno**, Sierra Detectors. 419 Flint. 89501. (702-323-2712).

NEW HAMPSHIRE: Concord, Don Wilson Sales. 93 So. State Street. 03301. (603-224-5909); **Seabrook**, The Village Trader. U.S. Route 1. 03874. (603-474-2836).

NEW JERSEY: Englewood, General Sales. #10 Humphrey Street. 07631. (201-568-5563); **Saddle Brook**, Geo-Quest. 104 US Hwy 46. 07662. (201-772-7443); **Trenton**, Treasure Cove. 1055 S. Clinton Avenue. 08611. (609-393-3631, 989-7382).

NEW MEXICO: Aztec, Wooley's Trailer Sales. 635 Aztec Boulevard. 87410. (505-334-2871); **Roswell**, Roswell Treasure Center. #12 Monterey Shopping Center. 1400 West Second Street. 88201. (505-623-2242).

NEW YORK: Fairport, Lost Coins Enterprise. Darrell C. Kilburn. 721 Mosley Rd. , 14450. (716-223-2139); **Geneva**, J. Panna's Electronic Sales. P.O. Box 167. 14456. (315-789-0809); **Glen Cove**, Fred Bond. 2 Leech Circle So. 11542. (516-676-1310); **Manhattan**, Louie Calamia. 54 East 8th Street. 10003. (212-254-1763); **New York City**, C-T Detectors. 4443 Murdock Avenue. 10466. (212-325-9582); **Walton**, Doc Dave's Treasure Finders. 54 Stockton Avenue. Route 206. 13856. (607-865-5188).

NORTH CAROLINA: Asheboro, Treasure World of North Carolina. East Dixie Drive. 27203. (919-629-6164); **Asheville**, Strings & Things. 1064 Patton Avenue. 28806. (704-258-3589); **Bladenboro**, Miller Electronics.

Route 2. Box 744. 28320. (919-866-5600); **Charlotte,** Ernie "Carolina" Curlee, Detector Sales Co., Division of Chemation. 3201 Cullman Avenue. 28206. (704-375-8468. 537-5115); **Glen Raven,** Barbee Detector Sales. c/o Barbee Fabrics, Inc., P.O. Box 4235. 27215. (919-584-7781, 584-7873); **Moncure,** B & R Detector Sales. Route 1, Box 185-D. 27559. (919-542-2210, 542-3832); **Wilmington,** Russ Simmons. 414 Biscayne Drive. 28405. (919-686-7009).

NORTH DAKOTA: Fargo, Treasure Island. West Acres Shopping Center. 58103. (701-282-4747); **Minot,** Chester N. Iverson. 808 17th Avenue S.W. 58701. (701-838-0149).

OHIO: Ashtabula, McCoy Electronic Repair. 7830 Sanborn Road. 44004. (216-997-4050); **Cincinnati,** J & B Treasures (Northwest Accessories). 2163 Sevenshills Drive. 45240. (513-742-3344); **Cleveland,** Kilian Detector Equipment Company. 1031 Spring Road. 44109. (216-398-4779); **Elyria,** T & K Cycles. 36668 Butternut Ridge Rd., 44035. (216-327-3783); **Lewisburg,** Fox Metal Detectors. Shields Road. RR #2. Box 312 D. 45338. (513-962-2937); **Lima,** Klingler's Rocks 'N Things. 1763 Bowman Road. 45804. (419-227-5294); **Millersport,** The Penny Place. P.O. Box 578. 43046. (614-467-2864); **Ottawa,** Winkle Radio & TV. Route 4. 17 Mi. N. Lima. 1½ Mi. N. Kalida. Route 115. 45875. (419-532-3957); **Pierpont,** G & D Detector Sales. 6500 North Richmond Road. 44082. (216-577-1496); **Shelby,** Struble Drug Inc. of Shelby. 31 West Main Street. 44875. (419-342-2136, 347-2802); **Toledo,** The Treasure House. 5734 Elmer. 43615. (419-531-7787); **Waterville,** The Treasure Chest. 3 Mi. W. Waterville. Route 24. 9204 S. River Road. 43566. (419-878-6026).

OKLAHOMA: Ada, Eddie S. Fausett Sales. 2729 Kirby Drive. 74820. (405-332-3156); **Jenks,** Woodrow J. Russey. 904 N. Juniper. 74037. (918-299-3551); **Keota,** James Bruner & Sons. Route One. 74941. (918-966-3779); **Oklahoma City,** Hobby World. 2433 Plaza Prom. Shepherd Mall. 73107. (405-942-4556); **Tulsa,** Ace's Detector Service. 5622 S. Pittsburg. 74135. (918-742-2214).

OREGON: Coos Bay, Carla Kay Salvage Co. 471 Brule Street. 97420. (503-888-4015); **Eugene,** Oregon Gold Dredge Limited. P.O. Box 10214. 50 Grimes Road. 97440. (503-686-2769); **Grant's Pass,** Kelly Metalcraft. 5057 Redwood Avenue. 97526. (503-479-1767); **Medford,** Medford Coin & Prospecting Supply. 411 East Main. 97501. (503-772-1477, 773-4471).

PENNSYLVANIA: Greensburg, Sealand's Metal Detectors. 422 Sells Lane. 15601. (412-834-3429); **Langhorne,** Sonny's Cycle & Sporting Arms. 1964 Maple Avenue. Route 213. 19047. (215-752-3030); **Meadville,** Miller's Treasure & Metal Detectors. RD #1. Pettis Rd. 16335. (814-336-5453); **Milford,** Warren Pedersen. Route 6. RR 2. Box 381. 18337. (717-296-7285); **Morrisdale,** Robert & Cleora Ferguson. P.O. Box 14. Route 53. 16858. (814-342-1268); **Williamsport,** K.A. Detectors. RD 4. Box 323. 17701. (717-326-0867).

RHODE ISLAND: Warwick, House of Bargains. 345 Warwick Avenue. 02888. (401-781-8580).

SOUTH CAROLINA: Sumter, Ken Lyles Detectors. 122 Lazy Lane. 29150. (home — 803-775-8840. office — 775-2806)

SOUTH DAKOTA: Rapid City, Donco Metal Detectors. 2424 Canyon Lake Drive. 57701. (605-343-3103)

TENNESSEE: Bolivar, West Tennessee Detector Sales. 322 Central Street. P.O. Box 162. 38008. (901-658-5196); **Chattanooga,** Chattanooga Detector Sales. 4710 3rd Avenue. 37407. (615-622-8882); **Chattanooga,** Hickory Valley Electric Co. and Metal Detector Sales. 6916 Lee Hwy.. 37421. (615-892-0525, 892-3581);

Memphis, Amoneco. 681 Madison. 38103. (901-526-5054); **Nashville,** The Collector's Shop. 100 Oaks Shopping Center. 37204. (615-383-5996); **Selmer,** Selmer Service Station & Sporting Goods. 100 West Court Avenue. 38375. (901-645-5431).

TEXAS: Amarillo, J. C. Claxton. 1602 Canyon Drive. 79102. 806-373-8971. 374-3820); **Austin,** Niles Carter. 2103 Whitestone Drive. 78745. (512-444-0106); **Beaumont,** Sanders Sports Shop. Route 8. Box G-36. 77705. (713-794-2560); **Brownsville,** Chester's Coin Shop. 2606 International Blvd., 78521. (512-546-4252); **Bryan,** Treasure Hound Detector Sales. 400 Mitchell. 77801. (713-779-6423. 845-2211); **Corpus Christi,** Bayside Metal Detectors. 9245 So. Padre Island Drive. 78418. (512-937-1682. 937-5334); **Dallas,** The Syndicate. 1066 Valley View Center. 75240. (214-233-6694); United Treasure Hunters. 11602 Garland Road. 75218. (214-328-1223); **Donna,** E. David Medrano. 410 So. 23rd Street. 78537. (512-464-4270); **El Paso,** American Camping & Outing Industries. Inc., P.O. Box 12564. 79912. (915-751-7741); **Ft. Worth,** Rex Grove Auto Supply Co., Inc., 4527 E. Belknap. 76117. (817-838-3066. 838-9640); **Granbury,** Roy's Detectors. Box 751. Hwy 377. Route 4. 76048. (817-573-9241); **Harlingen,** Green Gables — Metal Detector Sales. 1910 N. "77" Sunshine. 78550. (512-428-8420); **Houston,** Alexander Enterprises. 21 Spencer Highway. 77587. (713-946-6399); **Houston,** Research & Recovery. 2803 Old Spanish Trail. 77054. (713-747-4647, 747-4648); **Kountze,** Peterson's Trash & Treasure. P.O. Box 288. 77625. (713-246-2591); **Mission,** Mission Rexall Drug. 1030 Conway Avenue. 78572. (512-585-1532); **San Antonio,** Owens Detector Sales. 5814 Kepler Drive. 78228. (512-434-1605); **Uvalde,** Spurgeon's Artifacts & Coins. 205 W. Nueces. 78801. (512-278-2164); **Wichita Falls,** Eckhart's Detector Sales. 2503 Holliday Street. 76301. (817-767-3939).

UTAH: Bountiful, Pat Blackner's House of Treasure. 581 West 800 South. 84010. (801-292-3111); **Roy,** Bryant T. Cash. 2457 West 4975 South. 84067. (801-825-7858).

VIRGINIA: Fairfax, Suburban Detectors. 3169 Spring Street. 22030. (703-273-2542); **Richmond,** Essential Electronics. 10453 Medina Rd., 23235. (804-272-5558); **Virginia Beach,** H & S Detector Center. 2108 Thoroughgood Rd., 23455. (804-464-6072).

WASHINGTON: Auburn, Cache Inn Detectors. 17925 S.E. 313th. 98002. (206-631-0466); **Bellingham,** Washington Divers. 903 N. State Street. 98225. (206-676-8029); **Bremerton,** Tanner's Diggin's. 4029 Boundary Trail NW. 98310. (206-377-2532); **Kennewick,** The Coin Cradle Inc., 2810 W. Kennewick Avenue. Suite "E". 99336. (509-735-1507); **Seattle,** Pearl Electronics Inc., 1300 First Avenue. 98101. (206-622-6200); Prospector Ed's Gold Supplies. 5263 Rainier Avenue So., 98118. (206-723-8200); **Spokane,** Bowen's Hideout. S. 1823 Mt. Vernon. 99203. (509-534-4004).

WISCONSIN: Black River Falls, Ken's A.D.S., Route 1. Box 117. 54615. (715-284-2105); **Madison,** Pete's Rock Shop. 1917-19 Winnebago Street. 53704. (608-249-2648); **Sheboygan,** Jetzer's Locksmith Service. 3212 North 21st Street. 53081. (414-457-9231); **Waukesha,** Outdoor Outfitters. 705 Elm Ct., 53186. (414-542-7772).

WEST VIRGINIA: Paden City, Murdock's Hobby Shop. 121 N. Fourth Avenue. 26159. (304-337-2711); **Shady Spring,** Ray's Leisure Time Shop. P.O. Drawer E. US Highways. 19 & 21. 25918. (304-763-3110).

WYOMING: Casper, Caspar Metal Detectors Sales & Rentals. 1281 Payne and 1017 Cardiff. 82601. (307-235-6323, 234-5205).

FOREIGN

AUSTRALIA, Victoria, Park, P. J. Bridge Hesperian Detectors. P.O. Box 317. 6100. Western Australia. (09-32-57422, 32-58575.

CANADA:
ONTARIO, *Canadian Treasure Trail Ltd., P.O. Box 22, Camden East. K0K 1J0. (613-378-6421) *Distributor and Service Center for Canada; **Ayr,** Treasure Unlimited. Box 257. N0B 1E0. (519-632-7955); **Downsview,** Sub-Mariners Diving Equipment. 954 Wilson Avenue. M3K 1E7. (416-630-2590).

ALBERTA: Edmonton, Bedrock Detectors. 10250-82 Street. T6A 3M3. (403-469-3050); **Milk River,** Jerry's Detectors. P.O. Box 536. 508-4th Avenue N.E., T0K 1T0. (403-647-3851).

ONTARIO: Peterborough, Leisure Detector Sales. Box 44. K9J 6Y5. (705-745-7655).

ALBERTA: Rocky Mountain House, Discovery Detectors. Box 1284. T0M 1T0. (403-845-3718).

ONTARIO: Scarborough, Pirates Cove. 3274 Danforth Avenue. M1L 1C3. (416-691-5560); **Stirling,** Tall Pines Treasure Trail. Box 186. (613-395-2406); **Strathroy,** L. W. Electronics. Box 42. (519-245-1994).

BRITISH COLUMBIA, Vancouver, Diversified Electronics Limited. 1104 Franklin Street. V6A 1J6. (604-254-0761).

MANITOBA; Winnipeg, O.K. John, Stn. F. Box 54. R2L 2A5; (204-667-6556).

NEW BRUNSWICK; Stanley, York Carleton. Treasure Supplies. P.O. Box 147. E0H 1T0. (506-367-2955).

ONTARIO: Waterford, D. Keith Edwards. RR #5, N0E 1Y0. (519-443-5193).

SASKATCHEWAN: Fort Qu'Appelle, Ken Co Industries. P.O. Box 280. S0G 1S0. (332-5312).
Yorkton, John Menken. 67 Darlington Street E., S3N 0C4. (306-783-8336).

GREAT BRITAIN & IRELAND
ENGLAND: London, *Pieces of Eight. 259 Eversholt Street. N.W.1. (01-388-3686) *Distributor and Service Center for UK. **London,** Treasure World. 155 Robert Street. N.W.1. (01-387-3142).

MEXICO
CALIFORNIA: San Diego, Gem & Treasure Hunting Association. 2493 San Diego Avenue. 92110. (714-297-2672) (Closed Monday & Tuesday).

TEXAS: Donna, E. David Medrano. 410 So. 23rd Street. 78537. (512-464-4270).

PUERTO RICO: Caparra Terrace-Rio Piedras, Treasure Hunting Center. 1572 Jesus T. Pinero Avenue. 00921. (809-781-6902).

RECOMMENDED SUPPLEMENTARY BOOKS

The books described below are among the most popular books in print related to treasure hunting. If you desire to increase your skills in various aspects of treasure hunting, consider adding these volumes to your library.

DETECTOR OWNER'S FIELD MANUAL. Roy Lagal. Ram Publishing Company. Nowhere else will you find the detector operating instructions that Mr. Lagal has put into this book. He shows in detail how to treasure hunt, cache hunt, prospect, search for nuggets, black sand deposits... in short, how to use your detector exactly as it should be used. Covers completely BFO-TR-VLF/TR types, P.I.'s, P.R.G.'s, P.I.P's, etc. Explains precious metals, minerals, ground conditions, and gives proof that treasure exists because it has been found and that more exists that you can find! Fully illustrated. 236 pages. $6.95.

ELECTRONIC PROSPECTING. Charles Garrett, Bob Grant, Roy Lagal. Ram Publishing Company. A tremendous upswing in electronic prospecting for gold and other precious metals has recently occurred. High gold prices and unlimited capabilities of VLF/TR metal detectors have led to many fantastic discoveries. Gold is there to be found. If you have the desire to search for it and want to be successful, then this book will show you how to select (and use) from the many brands of VLF/TR's those that are correctly calibrated to produce accurate metal vs. mineral identification which is so vitally necessary in prospecting. Illustrated. 96 pages. $3.95.

GOLD PANNING IS EASY. Roy Lagal. Ram Publishing Company. Roy Lagal proves it! He doesn't introduce a new method; he removes confusion surrounding old established methods. A refreshing NEW LOOK guaranteed to produce results with the "Gravity Trap" or any other pan. Special metal detector instructions that show you how to nugget shoot, find gold and silver veins, and check ore samples for precious metal. This HOW, WHERE and WHEN gold panning book is a must for everyone, beginner or professional! Fully illustrated. 96 pages. $3.95.

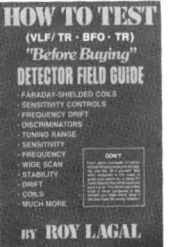

HOW TO TEST "BEFORE BUYING" DETECTOR FIELD GUIDE. Roy Lagal. Ram Publishing Company. Completely explains the inner workings of the BFO, TR, and discriminator types of detectors. You will learn how to test for sensitivity, stability, total response, wide scan, soil conditions, coils, Faraday shields, and frequency drift, and you will be able to expose incompetent detector engineering and overly enthusiastic, misleading advertising. If you own or are thinking of buying a detector, this book is an ABSOLUTE MUST. Fully illustrated. 64 pages. $3.95.

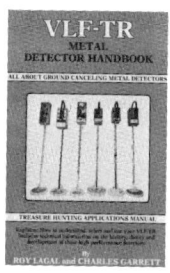
THE COMPLETE VLF-TR METAL DETECTOR HANDBOOK (All About Ground Canceling Metal Detectors). Roy Lagal, Charles Garrett. Ram Publishing Company. The unparalleled capabilities of VLF/TR Ground Canceling metal detectors have made them the number one choice of treasure hunters and prospectors. From History, Theory, and Development to Coin, Cache, and Relic Hunting, as well as Prospecting, the authors have explained in detail the capabilities of VLF/TR detectors and how they are used. Learn the new ground canceling detectors for the greatest possible success. Illustrated. 200 pages. $7.95.

THE JOURNALS OF EL DORADO. Estee Conatser, Karl von Mueller. Ram Publishing Company. A descriptive bibliography on treasure and related subjects; a first-of-its-kind storehouse of information devoted exclusively to information of interest to treasure hunters, prospectors, and relic hunters. This book contains approximately 1,800 book listings arranged alphabetically by author. It was developed as a working tool and reference for those in the treasure, small mining, and prospecting fields, especially beginners. Thousands of treasure leads will be found between its covers. Invaluable. 380 pages. $9.95.

TREASURE HUNTER'S MANUAL #6. Karl von Mueller. Ram Publishing Company. The original material in this book was written for the professional treasure hunter. Hundreds of copies were paid for in advance by professionals who knew the value of Karl's writing and wanted no delays in receiving their copies. The THM #6 completely describes full-time treasure hunting and explains the mysteries surrounding this intriguing and rewarding field of endeavor. You'll read this fascinating book several times. Each time you will discover you have gained greater in-depth knowledge. Thousands of ideas, tips, and other valuable information. Illustrated. 318 pages. $7.95.

TREASURE HUNTER'S MANUAL #7. Karl von Mueller. Ram Publishing Company. The classic! The most complete, up-to-date guide to America's fastest growing activity, written by the old master of treasure hunting. This is *the* book that fully describes professional methods of RESEARCH, RECOVERY, and TREASURE DISPOSITION. Includes a full range of treasure hunting methods from research techniques to detector operation, from legality to gold dredging. Don't worry that this material overlaps THM #6 . . . both of Karl's MANUALS are 100% different from each other but yet are crammed with information you should know about treasure hunting. Illustrated. 334 pages. $7.95.

SUCCESSFUL COIN HUNTING. Charles Garrett. Ram Publishing Company. The best and most complete guide to successful coin hunting, this book explains fully the how's, where's, and when's of searching for coins and related objects. It also includes a complete explanation of how to select and use the various types of coin hunting metal detectors. Based on more than twenty years of actual in-the-field experience by the author, this volume contains a great amount of practical coin hunting information that will not be found elsewhere. Profusely illustrated with over 100 photographs. 248 pages. $6.95.

TREASURE HUNTING PAYS OFF! Charles Garrett. Ram Publishing Company. This book will give you an excellent introduction to all facets of treasure hunting. It tells you how to begin and be successful in general treasure hunting; coin hunting; relic, cache, and bottle seeking; and prospecting. It describes the various kinds of metal/mineral detectors and tells you how to go about selecting the correct type for all kinds of searching. This is an excellent guidebook for the beginner, but yet contains tips and ideas for the experienced TH'er. Illustrated. 88 pages. $3.95.

THE COMPLETE BOOK OF COMPETITION TREASURE HUNTING. Ernie "Carolina" Curlee. Ram Publishing Company. This book gives the details you need to know to sponsor or compete successfully in an organized treasure hunt. All about everything from choosing a name for a hunt and promoting it to receiving the prize you may have won. Whole sections on "How To Sponsor" and "How To Win." Every metal detector owner/treasure hunter can benefit from Ernie's down-to-earth, plainly written information and instructions. A book that will pay for itself many times over! Fully illustrated. 88 pages. $5.95.

PROFESSIONAL TREASURE HUNTER. George Mroczkowski. Ram Publishing Company. Research is 90 percent of the success of any treasure hunting endeavor. You will become a better treasure hunter by learning how, through proper treasure hunting techniques and methods, George was able to find treasure sites, obtain permission to search (even from the U. S. Government), select and use the proper equipment, and then recover treasure in many instances. If treasure was not found, valuable clues and historical artifacts were located that made it worthwhile or kept the search alive. Profusely illustrated. 132 pages. $6.95.

SPECIAL PUBLICATIONS OF THE
International Treasure Hunting Society

The International Treasure Hunting Society (ITHS) publishes a quarterly journal, THE INTERNATIONAL TREASURE HUNTER. Each issue contains carefully selected "how to" information regarding treasure hunting, metal detecting, prospecting, relic hunting, and other projects, as well as the latest successful treasure hunting stories of treasures found world-wide. Other information about treasure-hunting clubs and competition treasure hunts is also included. Copies are distributed free to ITHS members. Non-members may purchase one or more publications from Ram Publishing Company for $1.50 each. Copies may also be ordered direct from ITHS, P.O. Box 3007, Garland, Texas 75041. Obtain information about ITHS by writing the same address or by calling 214-271-0800.

THE INTERNATIONAL TREASURE HUNTER, Vol. 1, No. 1. This premier issue contains complete information regarding the organization and founding of the ITHS. Several informative articles include, "Electronic Treasure Hunting: The First Fifty Years," "Europe: A Treasure Hunter's Paradise!," and "The Soldier's Legacy, Searching for Battlefield Relics." These are but three of the articles included in this first issue which has now become a collector's item. $1.50 each.

THE INTERNATIONAL TREASURE HUNTER, Vol. 2, No. 1. This second issue has numerous "how to" articles, including a special coin hunting article by famed treasure hunter T. R. Edds. Read this article to learn how to "Unlock Ocean Beach Treasure Vaults." Other special articles tell about gold hunting in Australia, the First International Championship Treasure Hunt, and the first treasure hunter to search the ghost town of Spring Creek, Colorado. There are many other articles. $1.50 each.

BOOK ORDER BLANK

See your detector dealer or bookstore or send check or money order directly to Ram for prompt, postage paid shipping, bookpost. If not completely satisfied return book(s) within 10 days for a full refund.

____ DETECTOR OWNER'S FIELD MANUAL **$6.95**

____ ELECTRONIC PROSPECTING **$3.95**

____ GOLD PANNING IS EASY **$3.95**

____ HOW TO TEST "BEFORE BUYING" DETECTOR FIELD GUIDE **$3.95**

____ COMPLETE VLF-TR METAL DETECTOR HANDBOOK (THE) (ALL ABOUT GROUND CANCELING METAL DETECTORS) **$7.95**

____ JOURNALS OF EL DORADO (THE) **$9.95**

____ TREASURE HUNTER'S MANUAL #6 **$7.95**

____ TREASURE HUNTER'S MANUAL #7 **$7.95**

____ SUCCESSFUL COIN HUNTING **$6.95**

____ TREASURE HUNTING PAYS OFF! **$3.95**

____ COMPLETE BOOK OF COMPETITION TREASURE HUNTING (THE) **$5.95**

____ PROFESSIONAL TREASURE HUNTER **$6.95**

____ INTERNATIONAL TREASURE HUNTER. Vol. 1, No. 1 **$1.50**

____ INTERNATIONAL TREASURE HUNTER. Vol. 2, No. 1 **$1.50**

Please add 35¢ for each book ordered (to a maximum of $1.00) for handling charges.

Total for Items $ _____

Texas Residents Add
5% State Tax   _____

Handling Charge   _____

Total of Above $ _____

ENCLOSED IS MY CHECK OR MONEY ORDER $ _____

NAME _____

ADDRESS _____

CITY _____

STATE _____ ZIP _____

☐ PLACE MY NAME ON YOUR MAILING LIST

Ram Publishing Company
P.O. Drawer 38649, Dallas, Texas 75238
Dept. HBK3
214-278-8439
DEALER INQUIRIES WELCOME